STUDY GUIDE BOOK

Special Effects Makeup

임희경 서경대학교 미용예술학과 교수

HUO TAO School of Film, Television & Communication,
Xiamen University of Technology

SEO KYEONG UNIVERSITY 서경대학교 글로벌 미용교재개발 연구회

청구문화사

Contens

Lesson 1

특수효과 분장
Special Effects Makeup

Lesson 1. 특수효과 분장 Special Effects Makeup

1. 특수효과 분장이란? What is Special Effects Makeup?

- 특수효과 분장은 영화, TV, 공연, 사진 촬영 등에서 사용되는 기술이다.
- 특수효과 분장은 캐릭터의 외모를 변화시키거나 현실에서는 불가능한 캐릭터의 이미지 효과를 만들어내기 위해 사용된다.
- 특수효과 분장은 상처, 변형된 얼굴, 화상, 노화 등의 효과 및 외계인, 괴물, 동물 등의 변신 효과를 표현할 수 있다.
- 특수효과 분장은 시청자가 스토리 전반의 이해하고 현실감 있는 시청할 수 있도록 도와 준다.
- 특수효과 분장은 배우가 본인 역할 캐릭터에 몰입할 수 있으며, 리얼리티reality를 살려 극 전반의 캐릭터 상황에 맞는 설정을 표현할 수 있다.
- 특수효과 분장의 종류는 각기 다른 효과를 만들어내기 위해 창의성과 기술적 숙련도 요구 및 다양한 재료가 사용된다.

이미지 삽입
Insert Image

2. 일반 분장 분류 Types of Regular Makeup

- 캐릭터의 특징에 따라 주로 기초화장품과 색조화장품 등을 사용하다.

일반 분장 종류 Types of Regular Makeup	설명 Description
스트레이트 분장 Straight Makeup	
내츄럴 분장 Natural Makeup	
웨딩 분장 Wedding Makeup	
파티 분장 Party Makeup	
섹시 분장 Sexy Makeup	
패션 분장 Fashion Makeup	
광고 분장 Advertising Makeup	
할로윈 분장 Halloween Makeup	
광대 분장 Clown Makeup	
노인 분장 Old Age Makeup	

3. 특수 분장 분류 Types of Special Effects Makeup

- 기초, 색조화장품, 라텍스, 실리콘, 핫폼, 콜드폼, 알콜베이스제품, 인조 피 등 다양한 재료를 사용한다.

특수 분장 분류 Classification of Special Effects Makeup	설명 Description
문신분장 Tattoo Makeup	
상처 분장 Wound Makeup	
화상분장 Burn Makeup	
질병분장 Disease Makeup	
동상분장 Frostbite Makeup	
노화 분장 Aging Makeup	
머리카락 및 수염 분장 Hair and Beard Effects	
변신 분장 Transformations Makeup	
프로스테틱 Prosthetics Makeup	
디지털 특수효과 Digital Special Effects Makeup	
특수 분장 Special Effects Makeup	

4. 장르별 특수분장 적용 사례1 Examples of Special Effects Makeup by Genre1

장르별 영화분장에 적용되는 분장의 종류를 모두 선택 하세요

① 스트레이트 분장 Straight Makeup
② 내츄럴 분장 Natural Makeup
③ 웨딩 분장 Wedding Makeup
④ 파티 분장 Party Makeup
⑤ 섹시 분장 Sexy Makeup
⑥ 패션 분장 Fashion Makeup
⑦ 광고 분장 Advertising Makeup
⑧ 할로윈 분장 Halloween Makeup
⑨ 광대 분장 Clown Makeup
⑩ 노인 분장 Old Age Makeup
⑪ 상처 분장 Wound Makeup
⑫ 화상 분장 Burn Makeup
⑬ 질병 분장 Disease Makeup
⑭ 동상 분장 Frostbite Makeup
⑮ 노화 분장 Aging Makeup
⑯ 수염 분장 Hair and Beard Effects
⑰ 변신 분장 Transformations Makeup
⑱ 프로스테틱 Prosthetics Makeup
⑲ 디지털 특수효과 Digital Special Effects Makeup
⑳ 특수효과 분장 Special Effects Makeup

1) 로맨스장르 영화분장 Makeup for Romance Genre Films

- 영화제목 Movie Title:
-
- 적용 분장 Applied Makeup:

이미지 삽입
Insert Image

2) 드라마장르 영화분장 Makeup for Drama Genre Films

- 영화제목 Movie Title:
-
- 적용 분장 Applied Makeup:
-

이미지 삽입
Insert Image

3) 코미디장르 영화분장 Makeup for Comedy Genre Films

- 영화제목 Movie Title:
-
- 적용 분장 Applied Makeup:
-

이미지 삽입
Insert Image

4. 장르별 특수분장 적용 사례 2 Examples of Special Effects Makeup by Genre 2

장르별 영화분장에 적용되는 분장의 종류를 모두 선택 하세요

① 스트레이트 분장 Straight Makeup
② 내츄럴 분장 Natural Makeup
③ 웨딩 분장 Wedding Makeup
④ 파티 분장 Party Makeup
⑤ 섹시 분장 Sexy Makeup
⑥ 패션 분장 Fashion Makeup
⑦ 광고 분장 Advertising Makeup
⑧ 할로윈 분장 Halloween Makeup
⑨ 광대 분장 Clown Makeup
⑩ 노인 분장 Old Age Makeup
⑪ 상처 분장 Wound Makeup
⑫ 화상 분장 Burn Makeup
⑬ 질병 분장 Disease Makeup
⑭ 동상 분장 Frostbite Makeup
⑮ 노화 분장 Aging Makeup
⑯ 수염 분장 Hair and Beard Effects
⑰ 변신 분장 Transformations Makeup
⑱ 프로스테틱 Prosthetics Makeup
⑲ 디지털 특수효과 Digital Special Effects Makeup
⑳ 특수효과 분장 Special Effects Makeup

4) 역사장르 영화분장 Makeup for Historical Genre Films

- 영화제목 Movie Title:
-
- 적용 분장 Applied Makeup:
-

이미지 삽입
Insert Image

---

5) 액션장르 영화분장 Makeup for Action Genre Films

- 영화제목 Movie Title:
-
- 적용 분장 Applied Makeup:
-

이미지 삽입
Insert Image

---

6) 전쟁장르 영화분장 Makeup for War Genre Films

- 영화제목 Movie Title:
-
- 적용 분장 Applied Makeup:
-

이미지 삽입
Insert Image

4. 장르별 특수분장 적용 사례 3 Examples of Special Effects Makeup by Genre 3

장르별 영화분장에 적용되는 분장의 종류를 모두 선택 하세요

① 스트레이트 분장 Straight Makeup
② 내추럴 분장 Natural Makeup
③ 웨딩 분장 Wedding Makeup
④ 파티 분장 Party Makeup
⑤ 섹시 분장 Sexy Makeup
⑥ 패션 분장 Fashion Makeup
⑦ 광고 분장 Advertising Makeup
⑧ 할로윈 분장 Halloween Makeup
⑨ 광대 분장 Clown Makeup
⑩ 노인 분장 Old Age Makeup
⑪ 상처 분장 Wound Makeup
⑫ 화상 분장 Burn Makeup
⑬ 질병 분장 Disease Makeup
⑭ 동상 분장 Frostbite Makeup
⑮ 노화 분장 Aging Makeup
⑯ 수염 분장 Hair and Beard Effects
⑰ 변신 분장 Transformations Makeup
⑱ 프로스테틱 Prosthetics Makeup
⑲ 디지털 특수효과 Digital Special Effects Makeup
⑳ 특수효과 분장 Special Effects Makeup

7) 범죄장르 영화분장 Makeup for Crime Genre Films

- 영화제목 Movie Title:
-
- 적용 분장 Applied Makeup:
-

이미지 삽입
Insert Image

8) 스릴러장르 영화분장 Makeup for Thriller Genre Films

- 영화제목 Movie Title:
-
- 적용 분장 Applied Makeup:
-

이미지 삽입
Insert Image

9) 공상과학장르 영화분장 Makeup for Science Fiction Genre Films

- 영화제목 Movie Title:
-
- 적용 분장 Applied Makeup:
-

이미지 삽입
Insert Image

4. 장르별 특수분장 적용 사례 4 Examples of Special Effects Makeup by Genre 4

장르별 영화분장에 적용되는 분장의 종류를 모두 선택 하세요

① 스트레이트 분장 Straight Makeup
② 내츄럴 분장 Natural Makeup
③ 웨딩 분장 Wedding Makeup
④ 파티 분장 Party Makeup
⑤ 섹시 분장 Sexy Makeup
⑥ 패션 분장 Fashion Makeup
⑦ 광고 분장 Advertising Makeup
⑧ 할로윈 분장 Halloween Makeup
⑨ 광대 분장 Clown Makeup
⑩ 노인 분장 Old Age Makeup
⑪ 상처 분장 Wound Makeup
⑫ 화상 분장 Burn Makeup
⑬ 질병 분장 Disease Makeup
⑭ 동상 분장 Frostbite Makeup
⑮ 노화 분장 Aging Makeup
⑯ 수염 분장 Hair and Beard Effects
⑰ 변신 분장 Transformations Makeup
⑱ 프로스테틱 Prosthetics Makeup
⑲ 디지털 특수효과 Digital Special Effects Makeup
⑳ 특수효과 분장 Special Effects Makeup

10) 판타지장르 영화분장 Makeup for Fantasy Genre Films

- 영화제목 Movie Title:
-
- 적용 분장 Applied Makeup:
-

이미지 삽입
Insert Image

11) 공포장르 영화분장 Makeup for Horror Genre Films

- 영화제목 Movie Title:
-
- 적용 분장 Applied Makeup:
-

이미지 삽입
Insert Image

12) 좀비장르 영화분장 Makeup for Zombie Genre Films

- 영화제목 Movie Title:
-
- 적용 분장 Applied Makeup:
-

이미지 삽입
Insert Image

Lesson 2

특수 효과 분장 기획
Special Effects Makeup Plan

Lesson 2. 특수 효과 분장 기획 Special Effects Makeup Plan

1. 특수 분장 기획 Special Effects Makeup Plan

- 아이디어 Idea

- 내용 Content

이미지 삽입
Insert Image

2. 특수 분장 기획 적용 선택 Selection of Special Effects Makeup Planning and Application

특수 분장 분류 Classification of Special Effects Makeup	적용 항목 설명 Explanation of Applied Items
프로스테틱 Prosthetics Makeup	
상처 분장 Wound Makeup	
노화 Aging Makeup	
머리 및 수염 분장 Hair and Beard Effects	
변신 Transformations Makeup	
특수 분장 Special Effects Makeup	
디지털 특수효과 Digital Special Effects Makeup	
특수 분장 활용 사업 Special Effects Makeup Business	

3. 특수 분장 기획 적용 사례 Examples of Special Effects Makeup Planning and Application

① 역사장르 영화분장
Makeup in Historical Genre Films

② 액션장르 영화분장
Makeup in Action Genre Films

③ 공상과학장르 영화분장
Makeup in Science Fiction Genre Films

④ 공포장르 영화분장
Makeup in Horror Genre Films

⑤ 판타지장르 영화분장
Makeup in Fantasy Genre Films

⑥ 좀비장르 영화분장
Makeup in Zombie Genre Films

1) 모티브 장르 영화 분장 Makeup Inspired by Genre Films

☐ 영화 제목 Film Title

이미지 삽입 Insert Image	이미지 삽입 Insert Image

2) 모티브 장르 영화 분장 Makeup Inspired by Genre Films

☐ 영화 제목 Film Title

이미지 삽입 Insert Image	이미지 삽입 Insert Image

3) 모티브 장르 영화 분장 Makeup Inspired by Genre Films

☐ 영화 제목 Film Title

이미지 삽입 Insert Image	이미지 삽입 Insert Image

4. 작품 주제 선정 Selection of Work Theme

1) 작품 주제 선정 조별 회의 Group Meeting for Theme Selection

작품 주제 Subject of Work	영화 장르 Film Genre	영화 제목 Film Title	
		좋은 캐릭터 Heroic Character	악당 캐릭터 Villain Character
주제 Subject			
주제 Subject			
주제 Subject			
주제 Subject			
주제 Subject			

2) 작품주제 방향 설정 역할 분담1 Theme and Direction of Work Role Assignment1

NO	역할 Role	성명 Name	내용 Content
1	팀 장 Team Leader		
2	전시 계획 Exhibition Plan		
3	작품 설치1 Artwork Installation 1		
4	작품 설치2 Artwork Installation 2		
5	작품 철수1 Artwork Dismantling 1		
6	작품 철수2 Artwork Dismantling 2		

2) 작품주제 방향 설정 역할 분담2 Theme and Direction of Work Role Assignment2

NO	역할 Role	성명 Name	내용 Content
7	총 무 Treasurer		
8	도록 제작-이미지 Catalog Production -Image		
9	도록 제작-내용 Catalog Production -Content		
10	사진 촬영 Photography		
11	영상 편집1 Video Editing 1		
12	영상 편집 2 Video Editing 2		
13			

5. 특수 분장 기획 적용 사례 Examples of Special Effects Makeup Planning and Application

① 역사장르 영화분장.
Makeup in Historical Genre Films.

② 액션장르 영화분장.
Makeup in Action Genre Films.

③ 공상과학장르 영화분장.
Makeup in Science Fiction Genre Films.

④ 공포장르 영화분장.
Makeup in Horror Genre Films.

⑤ 판타지장르 영화분장.
Makeup in Fantasy Genre Films

⑥ 좀비장르 영화분장.
Makeup in Zombie Genre Films

1) 모티브 캐릭터 Inspired Character

□ 영화 제목 Film Title

□ 1안: 모티브 캐릭터명 Option 1: Name of Inspired Character

이미지 삽입 Insert Image	이미지 삽입 Insert Image

□ 2안: 모티브 캐릭터명 Option 2: Name of Inspired Character

이미지 삽입 Insert Image	이미지 삽입 Insert Image

□ 3안: 모티브 캐릭터명 Option 3: Name of Inspired Character

이미지 삽입 Insert Image	이미지 삽입 Insert Image

6. 스토리 및 캐릭터 분석 Story and Character Analysis

1) 스토리 작성 Story Writing

2) 시대, 배경 분석 Era and Background Analysis

이미지 삽입
Insert Image

3) 캐릭터 분장 의도 Character Makeup Intent

4) 등장인물 분석 Character Analysis

5) 상황 분석 Situation Analysis

Lesson 3

특수효과 분장 재료
Special Effects Makeup Materials

Lesson 3. 특수효과 분장 재료 Special Effects Makeup Materials

1. 조소 도구 Sculpting Tools

도구 Tool	사용방법 How to use
도구 이미지 Tool Images	
도구 이미지 Tool Images	
도구 이미지 Tool Images	
도구 이미지 Tool Images	
도구 이미지 Tool Images	
도구 이미지 Tool Images	

2. 특수 분장 장비 Special Effects Makeup Equipment

장비 Equipment	사용방법 How to use
장비 이미지 Equipment Images	
장비 이미지 Equipment Images	
장비 이미지 Equipment Images	
장비 이미지 Equipment Images	
장비 이미지 Equipment Images	

3. 몰드제작 재료 Mold Making Materials

1) 석고 Plaster

이미지 삽입
Insert Image

2) 석고붕대 Plaster Bandages

이미지 삽입
Insert Image

3) 알지네이트 Alginate

이미지 삽입
Insert Image

4) 기타 Miscellaneous

이미지 삽입
Insert Image

4. 폼 라텍스 Foam latex

1) 라텍스 베이스 Latex Base

2) 포밍제 Foaming Agent

3) 양생제 Curing Agent

4) 겔링제 Gelling Agent

5) 분리제 Mold Release

이미지 삽입
Insert Image

5. 콜드 폼 Cold Foam

1) 폼 Foam

2) 경화제 Hardener

3) 기타 Miscellaneous

이미지 삽입
Insert Image

6. 실리콘 Silicon

1) 축합형 실리콘 Condensation Cure Silicone

2) 주가형 실리콘 Addition Cure Silicone

3) 몰드용 실리콘 Mold-Making Silicone

4) 첨가제 Additive

5) 경화촉진제 Accelerator

이미지 삽입
Insert Image

7. 접착제 Adhesive

이미지 삽입
Insert Image

이미지 삽입
Insert Image

이미지 삽입
Insert Image

8. 채색 재료 Coloring Materials

1)

이미지 삽입
Insert Image

2)

이미지 삽입
Insert Image

3)

이미지 삽입
Insert Image

4)

이미지 삽입
Insert Image

Lesson 4

캐릭터 분석
Character Analysis

Lesson 4. 캐릭터 분석 Character analysis

1. 캐릭터 자료 조사 Character Research

1) 자료조사 Research

이미지 삽입
Insert Image

2) 자료조사 Research

•

모티브 이미지 삽입
Insert Motif Image

•

모티브 이미지 삽입
Insert Motif Image

•

모티브 이미지 삽입
Insert Motif Image

•

모티브 이미지 삽입
Insert Motif Image

2. 캐릭터 디자인 개발 프로세스 Character Design Development Process

1) 랜덤워드 Random word

- 주제와 관련 없는 무작위 단어 선택

① **무작위 단어 선택**
Random Word Selection
② **연상 단어**
Associated Word
③ **연상 의미**
Meaning of Association
④ **주제 연결**
Topic Connection
⑤ **아이디어 도출**
Idea Draw

이미지 삽입
Insert Image

---

2) 마인드 맵 Mind Map

- 일을 시행하기전 계획을 그려 보는 것

① **공감하기**
Empathize
② **문제 정의하기**
Define
③ **아이디어 도출**
Ideating
④ **시제품 만들기**
Prototype
⑤ **테스트 하기**
Test

이미지 삽입
Insert Image

3) 브레인 라이팅 Brainwriting

- 생각나는 아이디어를 글로 작성 후 아디디어 공유 및 추가

① **개별 주제 선정**
Select Individual Topics
② **생각나는 아이디어 작성**
Write an Idea that Comes to Mind
③ **아이디어 공유**
Share Ideas
④ **아이디어 추가**
Add Ideas

☐
☐
☐
☐
☐

이미지 삽입
Insert Image

---

4) 디자인싱킹 Design Thinking

- 떠오르는 생각을 여과 없이 쏟아냄

① **공감하기**
Empathize
② **문제 정의하기**
Define
③ **아이디어 도출**
Ideating
④ **프로토타입 제작**
Prototype
⑤ **테스트**
Test

☐
☐
☐
☐
☐

이미지 삽입
Insert Image

5) 브레인스토밍 Brainstorming

- 떠오르는 생각을 여과 없이 쏟아냄

① 비판 금지
No Criticism
② 자유 분방
Freewheeling
③ 질보다는 양
Sheep Rather than Quality
④ 결합과 개선
Combining and Improving

☐
☐
☐
☐
☐
☐
☐
☐
☐
☐
☐
☐
☐
☐
☐

이미지 삽입
Insert Image

3. 캐릭터 분석 생성형 AI 활용 Utilizing Generative AI for Character Analysis

1) 생성형 AI Chat GPT 활용 Utilizing Generative AI Chat GPT

2) 이미지 생성형 AI 활용 Utilizing Image Generative AI

모티브 이미지 삽입 Insert Motif Image	모티브 이미지 삽입 Insert Motif Image

4. 분장 디자인 러프 스케치 Makeup Design Rough Sketch

1안 정면 1 Draft Front View

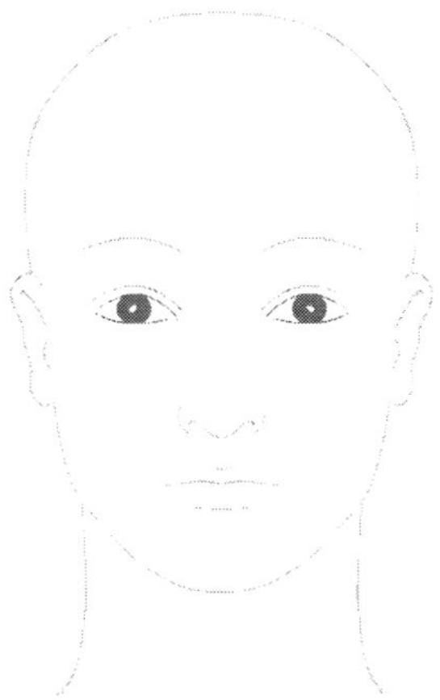

2안 정면 2Draft Front View

1안 측면 1 Draft Side View

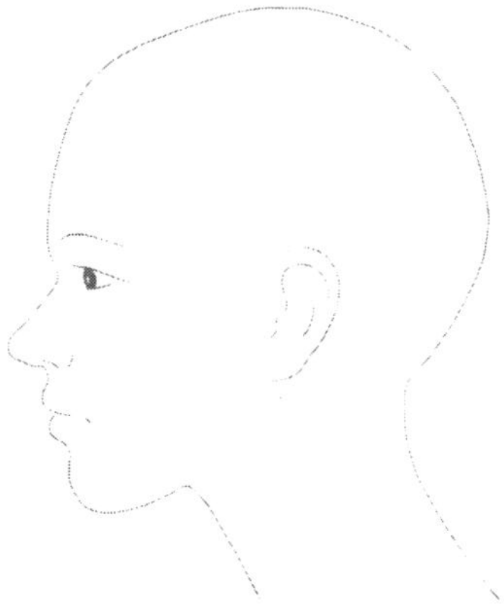

2안 측면 2 Draft Side View

전신 1 Full Body 1

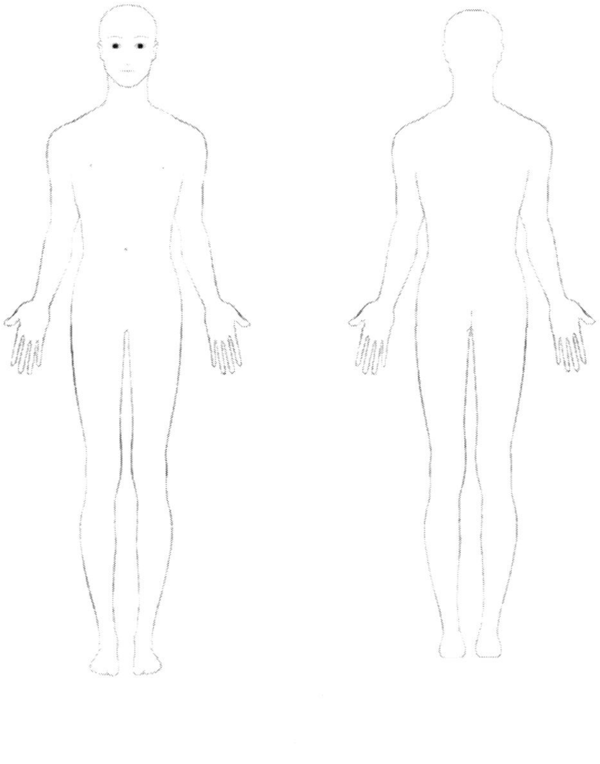

기타 Miscellaneous

5. 분장 디자인 - 정면 Makeup Design-Front

캐릭터의 정면 얼굴을 디자인하여 일러스트레이션하시오.

Design the front face of the character and create an illustration.

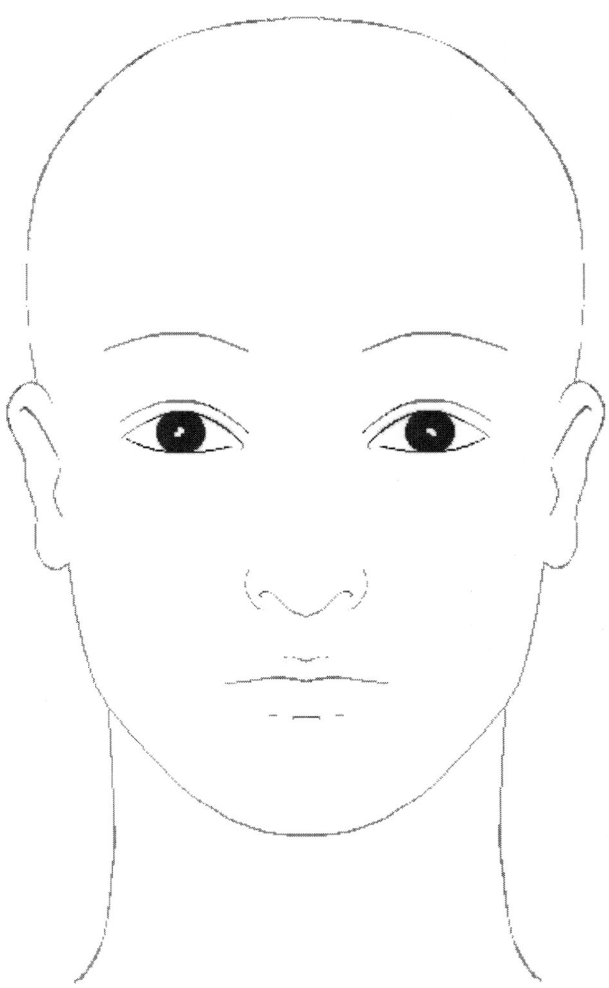

5. 분장 디자인 - 측면 Makeup Design-Side

캐릭터의 측면 얼굴을 디자인하여 일러스트레이션하시오.
Design and illustrate the side face of the character.

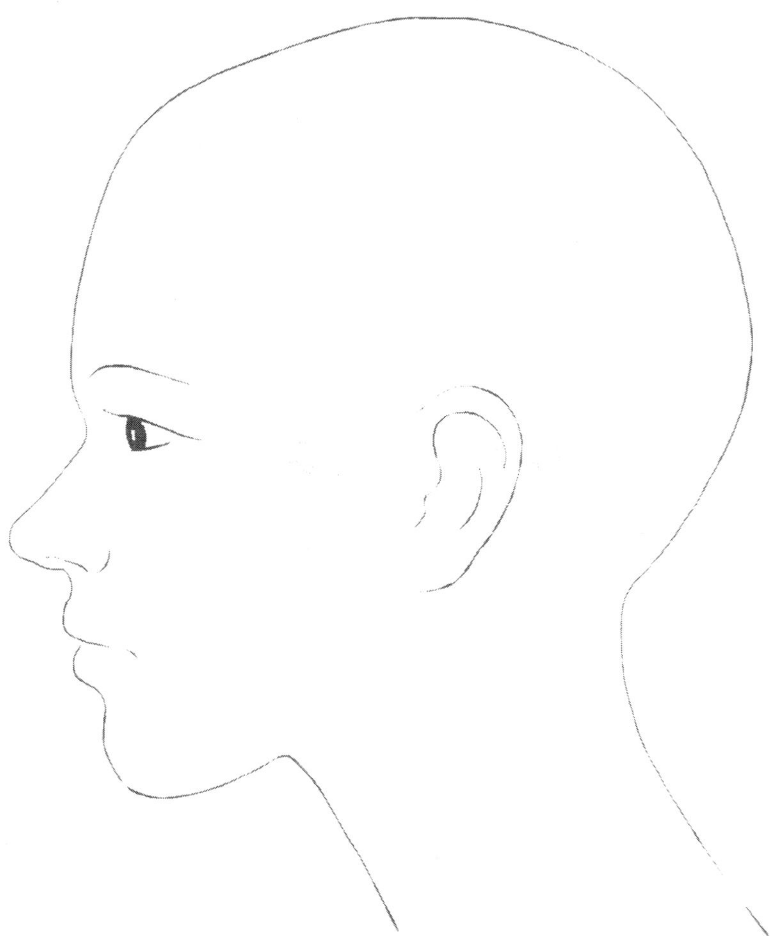

5. 분장 디자인 - 전신 Makeup Design - Full Body

캐릭터 전신의 분장 디자인을 일러스트레이션하시오.
Illustration of the makeup design of the character's entire body.

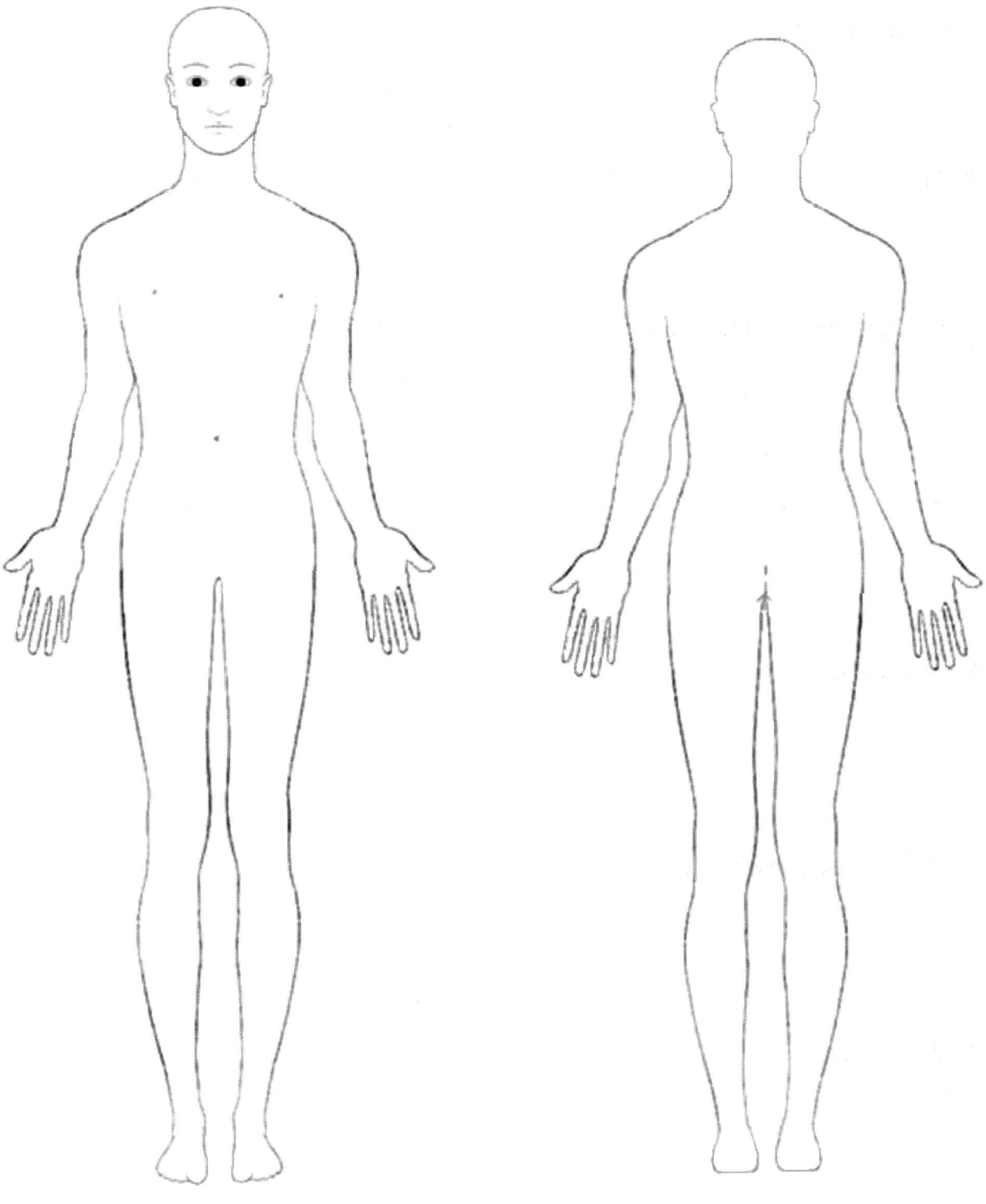

6. 캐릭터 분석 Character Analysis

<table>
<tr><th colspan="5">캐릭터 분석 Character Analysis</th></tr>
<tr><td colspan="2">작품명 Work Name</td><td colspan="2">장르 Genre</td><td>국가 Nationality</td></tr>
<tr><td colspan="2">배역명 Character Name</td><td colspan="3">역할 Part</td></tr>
<tr><td>성별 Sex</td><td colspan="2">나이 Age</td><td colspan="2">성격 Personality</td></tr>
<tr><td colspan="2">유전학적 특징 Genetic Characteristics</td><td colspan="3">시대적 배경 Historical Background</td></tr>
<tr><td colspan="2">건강 상태 Health Status</td><td colspan="3">생활 환경 Living Conditions</td></tr>
<tr><td colspan="5">특징 Characteristic</td></tr>
<tr><td colspan="5">성격 표현 Character Expression</td></tr>
<tr><td colspan="5">분장 의도 Makeup Purpose</td></tr>
</table>

7. 분장 계획표 Makeup Planning Chart

분장 계획표 Makeup Planning Chart

작품 명 Work Name	배역 명 Character Name
	배우 명 Actor's Name

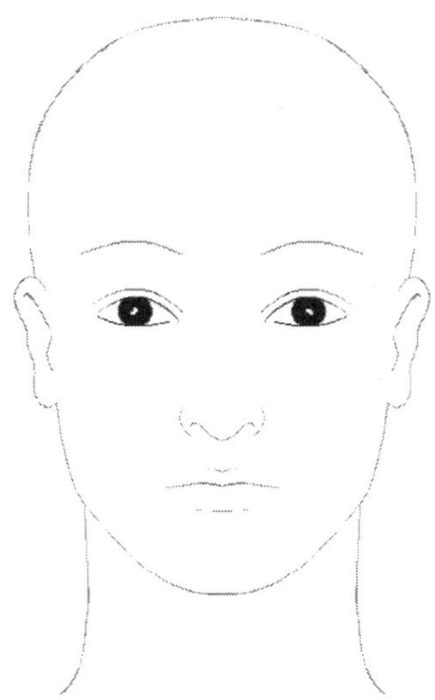

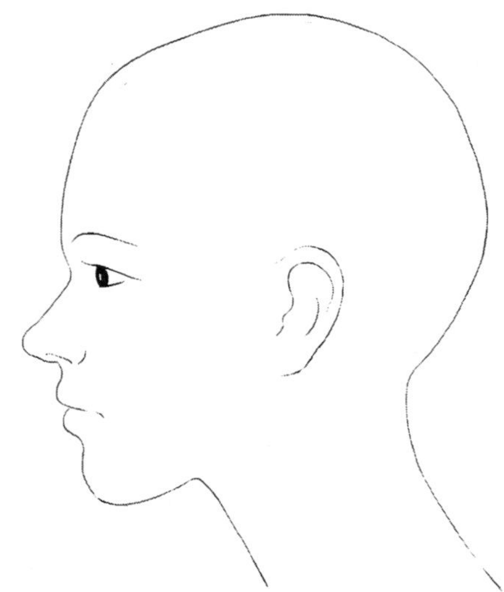

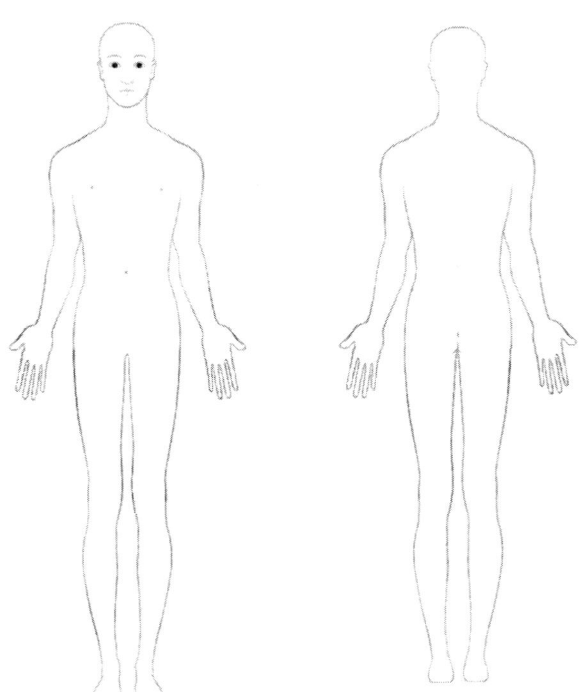

기초 Base	화운데이션 Foundation	음영 Contour
파우더 Powder	이마 Forehead	눈썹 Eyebrows
아이섀도 Eye shadow	아이라이너 Eye liners	마스카라 Mascara
코 Nose	뺨 Cheek	입 Mouth
헤어 Hair	수염 Beard	메모 Note

8. 캐릭터 분장 완성 Character Makeup Complete

분장 완성 이미지 삽입
Attached is an image of the Makeup completion.

Lesson 5

캐릭터 디자인 1
Character Design 1

Lesson 5. 캐릭터 디자인1 Character Design 1

1. 캐릭터 자료 조사 Character Research

이미지 삽입
Insert Image

2. 캐릭터 디자인 모티브 Character Design Motif

-
-

모티브 이미지 삽입
Insert Motif Image

모티브 이미지 삽입
Insert Motif Image

-
-

모티브 이미지 삽입
Insert Motif Image

모티브 이미지 삽입
Insert Motif Image

3. 분장 디자인 러프 스케치 Makeup Design Rough Sketch

1안 정면 1Draft Front View

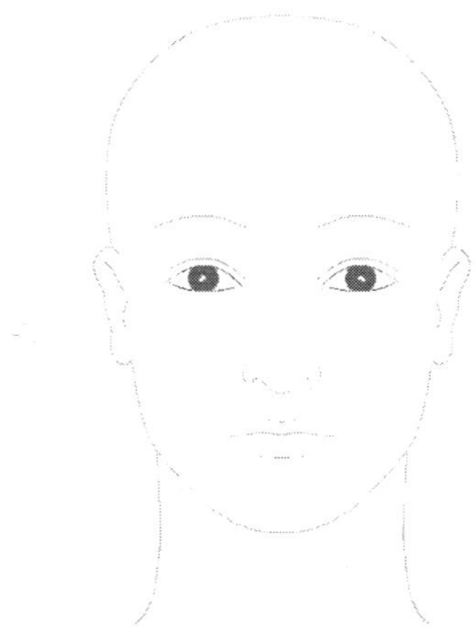

2안 정면 2Draft Front View

1안 측면 1 Draft Side View

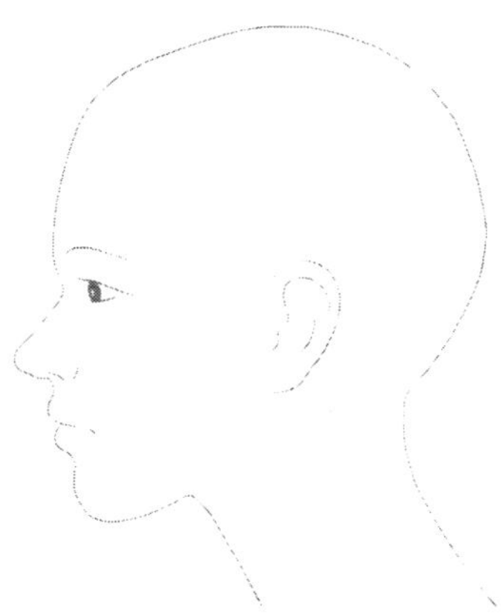

2안 측면 2 Draft Side View

전신 1 Full Body 1

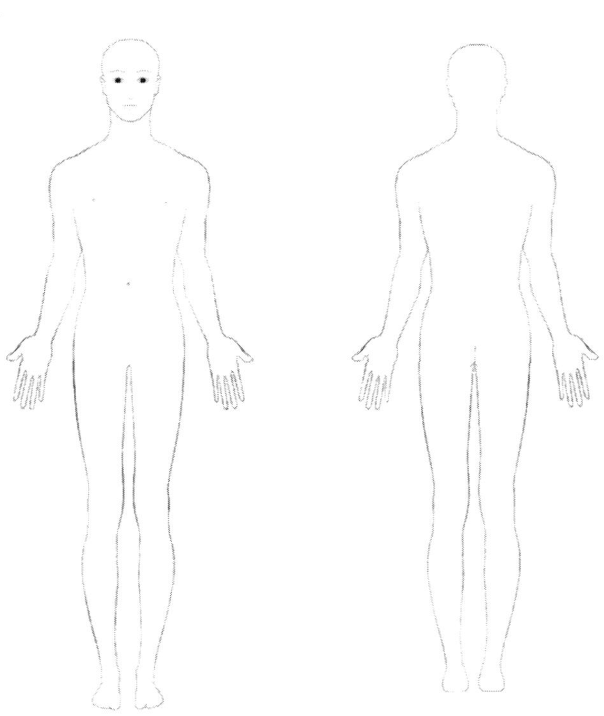

기타 Miscellaneous

4. 분장 디자인 - 정면 Makeup Design - Front

캐릭터의 정면 얼굴을 디자인하여 일러스트레이션하시오.
Design the front face of the character and create an illustration.

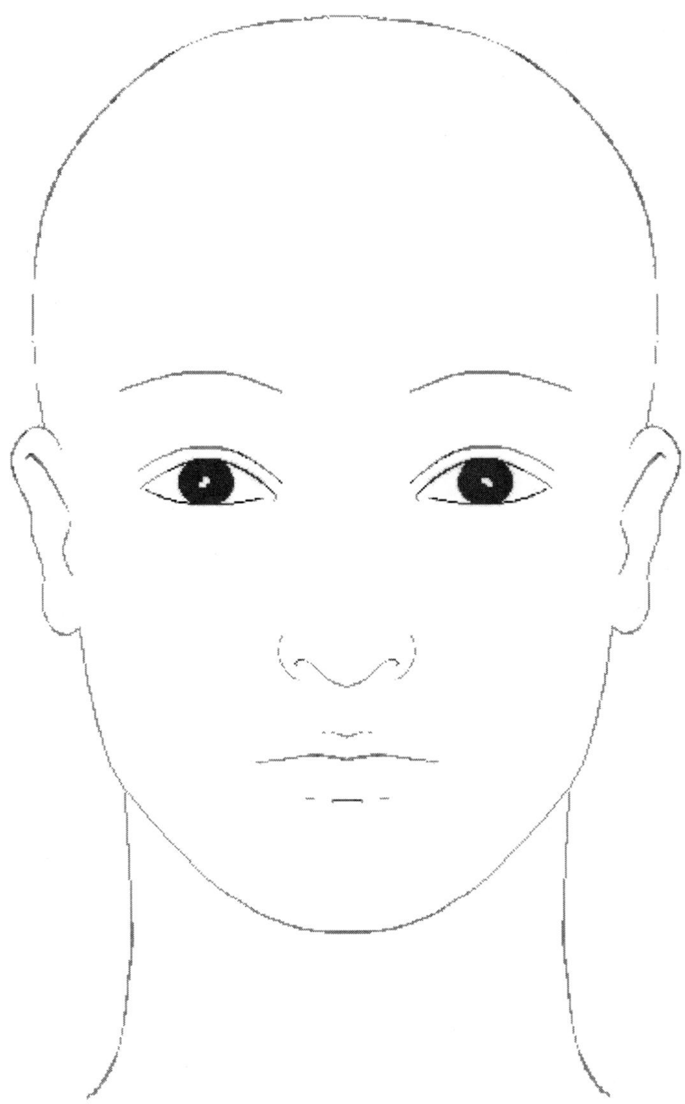

4. 분장 디자인 - 측면 Makeup Design - Side

캐릭터의 측면 얼굴을 디자인하여 일러스트레이션하시오.
Design and illustrate the side face of the character.

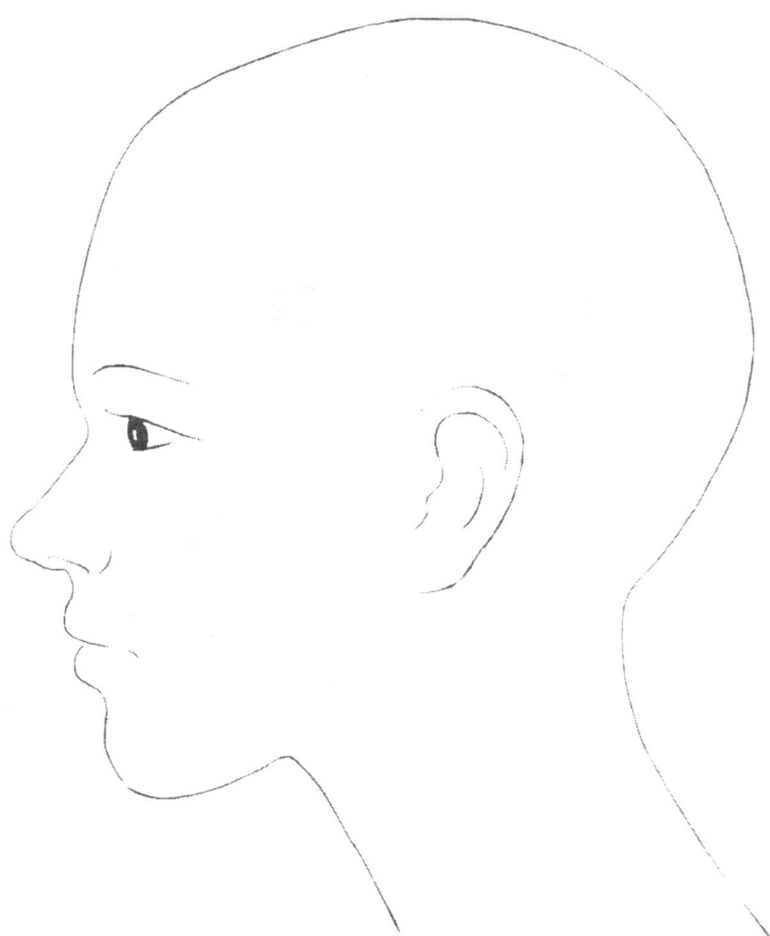

4. 분장 디자인 - 전신 Makeup Design - Full Body

캐릭터 전신의 분장 디자인을 일러스트레이션하시오.
Illustration of the makeup design of the character's entire body.

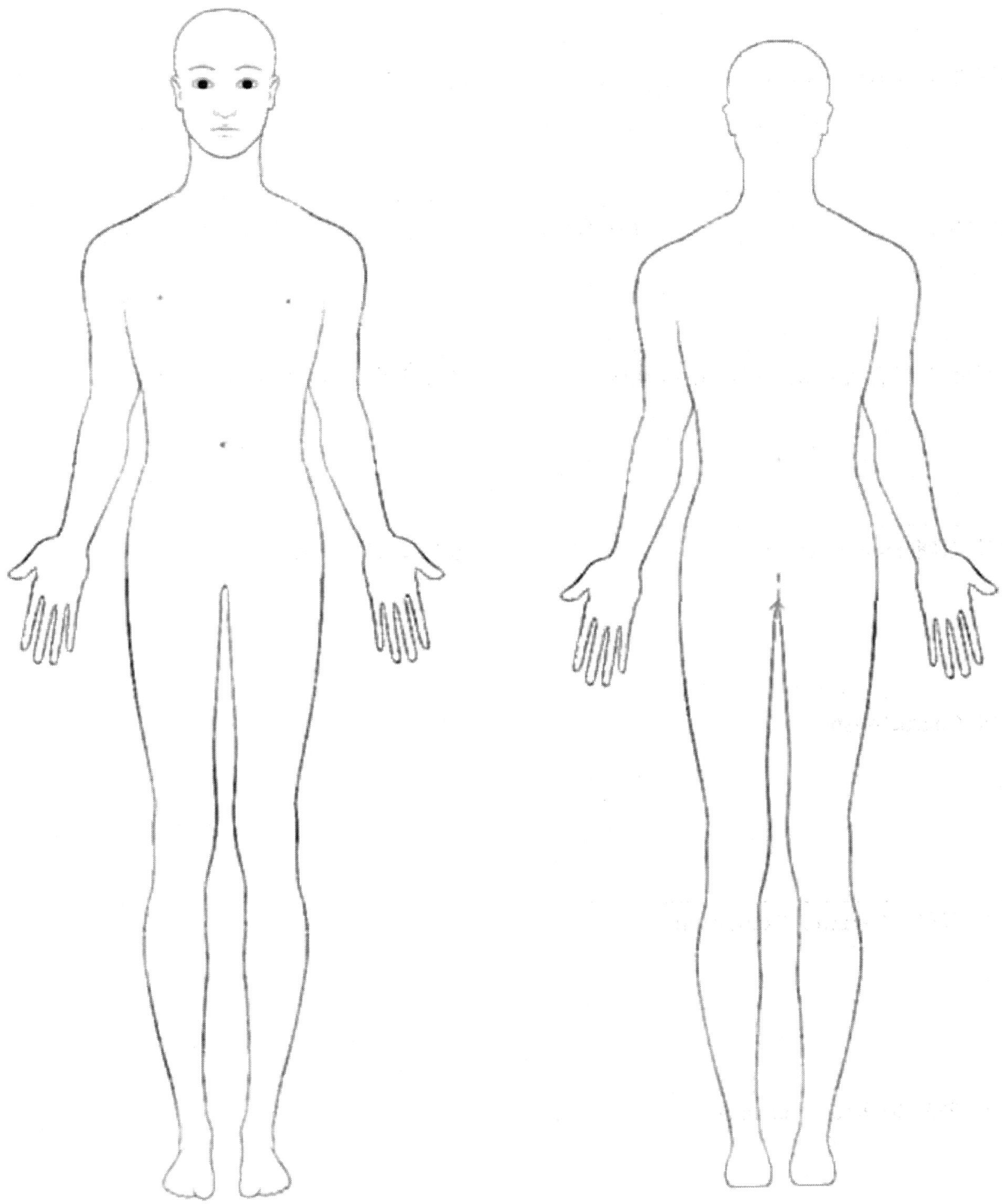

5. 캐릭터 분석 Character Analysis

<table>
<tr><th colspan="6">캐릭터 분석 Character Analysis</th></tr>
<tr><td colspan="3">작품명 Work Name</td><td colspan="2">장르 Genre</td><td>국가 Nationality</td></tr>
<tr><td colspan="3">배역명 Character Name</td><td colspan="3">역할 Part</td></tr>
<tr><td colspan="2">성별 Sex</td><td colspan="2">나이 Age</td><td colspan="2">성격 Personality</td></tr>
<tr><td colspan="3">유전학적 특징 Genetic Characteristics</td><td colspan="3">시대적 배경 Historical Background</td></tr>
<tr><td colspan="3">건강 상태 Health Status</td><td colspan="3">생활 환경 Living Conditions</td></tr>
<tr><td colspan="6">특징 Characteristic</td></tr>
<tr><td colspan="6">성격 표현 Character Expression</td></tr>
<tr><td colspan="6">분장 의도 Makeup Purpose</td></tr>
</table>

6. 분장 계획표 Makeup Planning Chart

분장 계획표 Makeup Planning Chart

작품 명 Work Name	배역 명 Character Name
	배우 명 Actor's Name

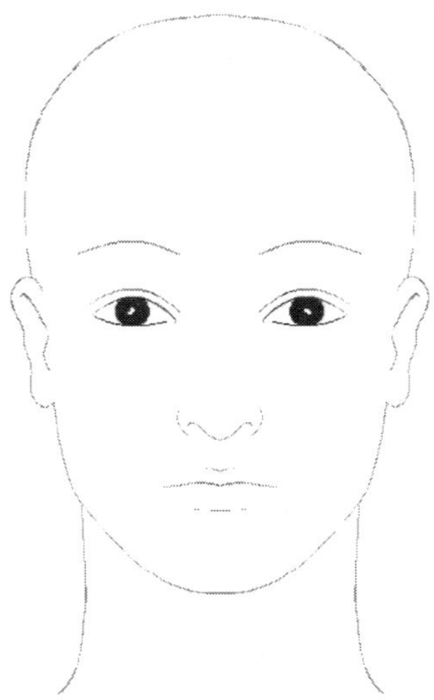

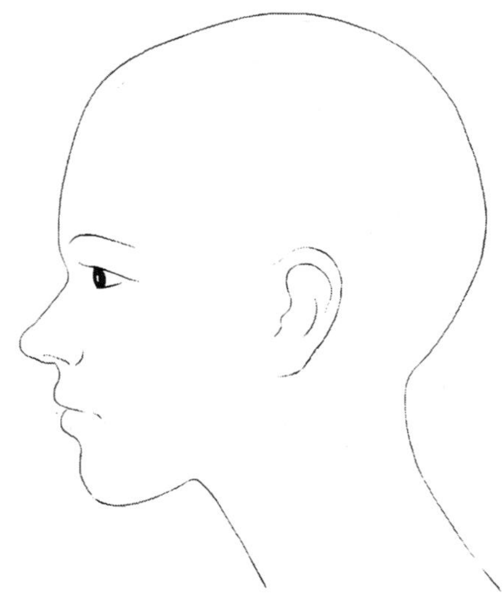

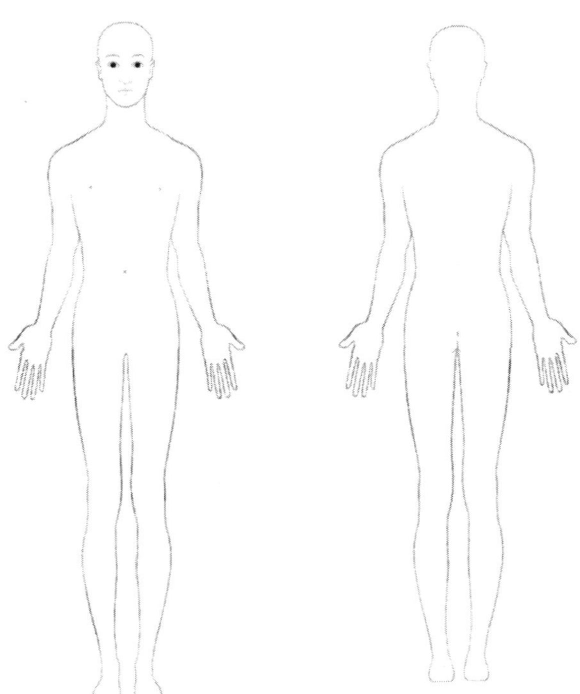

기초 Base	화운데이션 Foundation	음영 Contour
파우더 Powder	이마 Forehead	눈썹 Eyebrows
아이섀도 Eye shadow	아이라이너 Eye liners	마스카라 Mascara
코 Nose	뺨 Cheek	입 Mouth
헤어 Hair	수염 Beard	메모 Note

7. 캐릭터 분장 완성 Character Makeup Complete

분장 완성 이미지 삽입

Attached is an image of the Makeup completion.

Lesson 6

캐릭터 디자인 2

Character Design 2

Lesson 6. 캐릭터 디자인2 Character Design 2

1. 캐릭터 자료 조사 Character Research

이미지 삽입
Insert Image

2. 캐릭터 디자인 모티브 Character Design Motif

-
-

모티브 이미지 삽입
Insert Motif Image

모티브 이미지 삽입
Insert Motif Image

-
-

모티브 이미지 삽입
Insert Motif Image

모티브 이미지 삽입
Insert Motif Image

3. 분장 디자인 러프 스케치 Makeup Design Rough Sketch

1안 정면 1Draft Front View

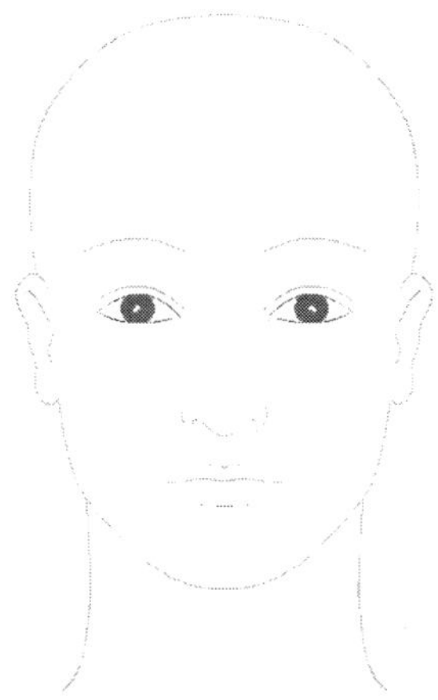

2안 정면 2Draft Front View

1안 측면 1 Draft Side View

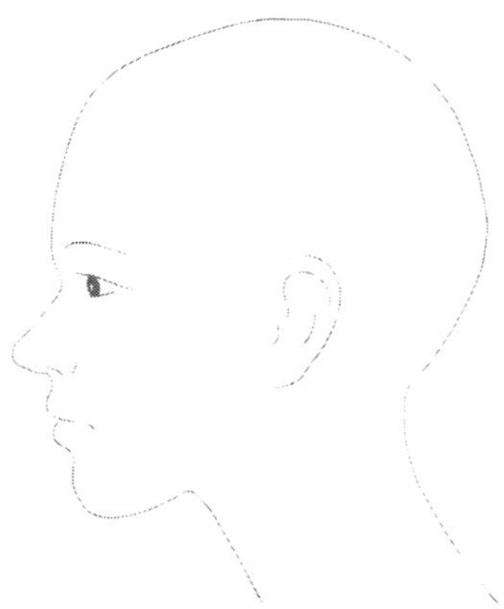

2안 측면 2 Draft Side View

전신 1 Full Body 1

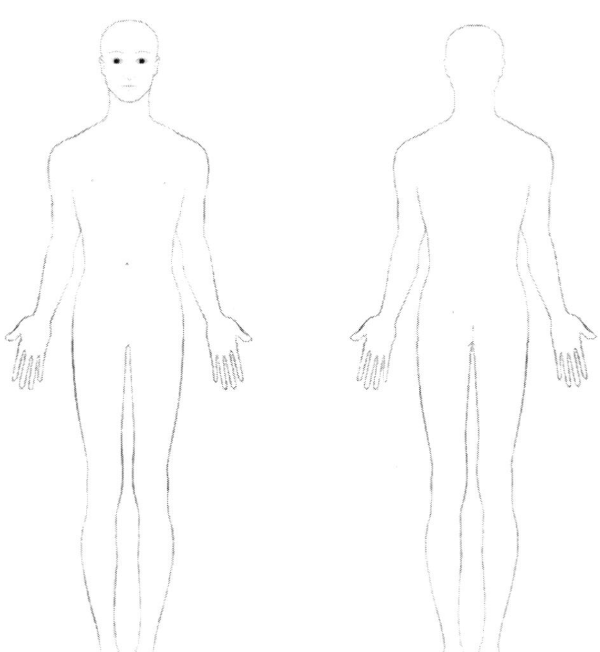

기타 Miscellaneous

4. 분장 디자인 - 정면 Makeup Design - Front

캐릭터의 정면 얼굴을 디자인하여 일러스트레이션하시오.

Design the front face of the character and create an illustration.

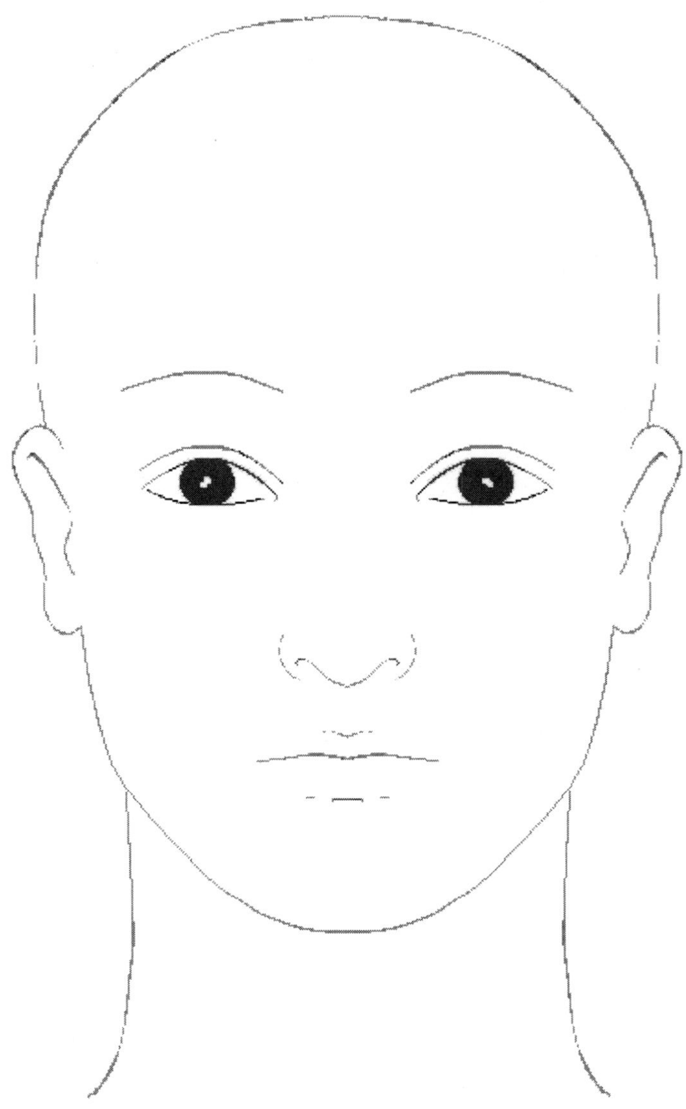

4. 분장 디자인 - 측면 Makeup Design - Side

캐릭터의 측면 얼굴을 디자인하여 일러스트레이션하시오

Design and illustrate the side face of the character.

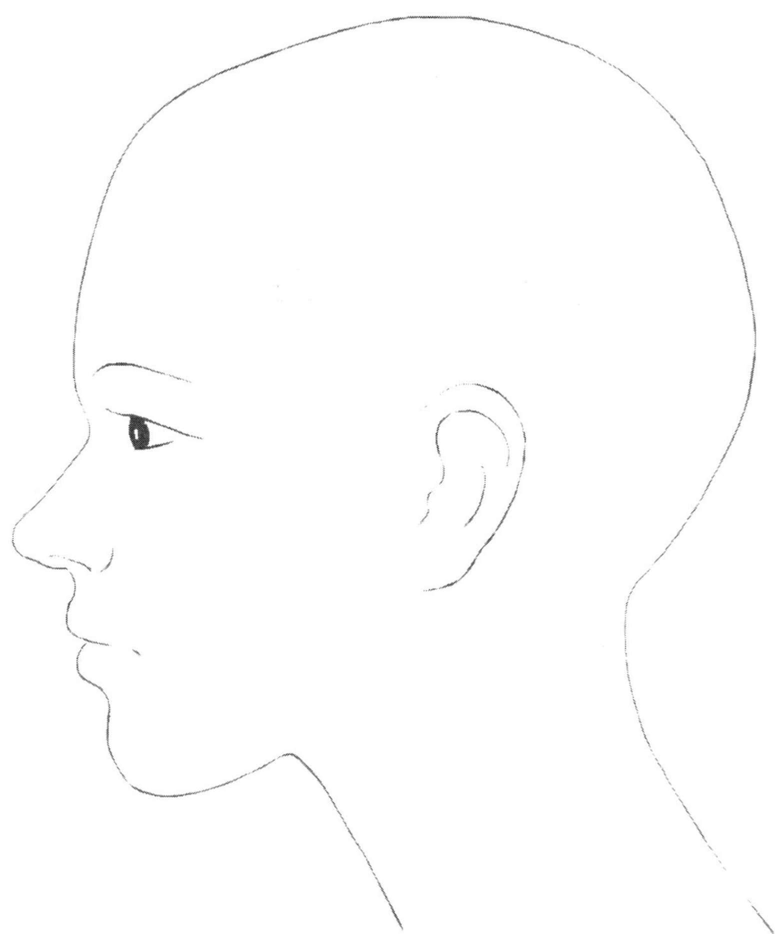

4. 분장 디자인 - 전신 Makeup Design - Full Body

캐릭터 전신의 분장 디자인을 일러스트레이션하시오.
Illustration of the makeup design of the character's entire body.

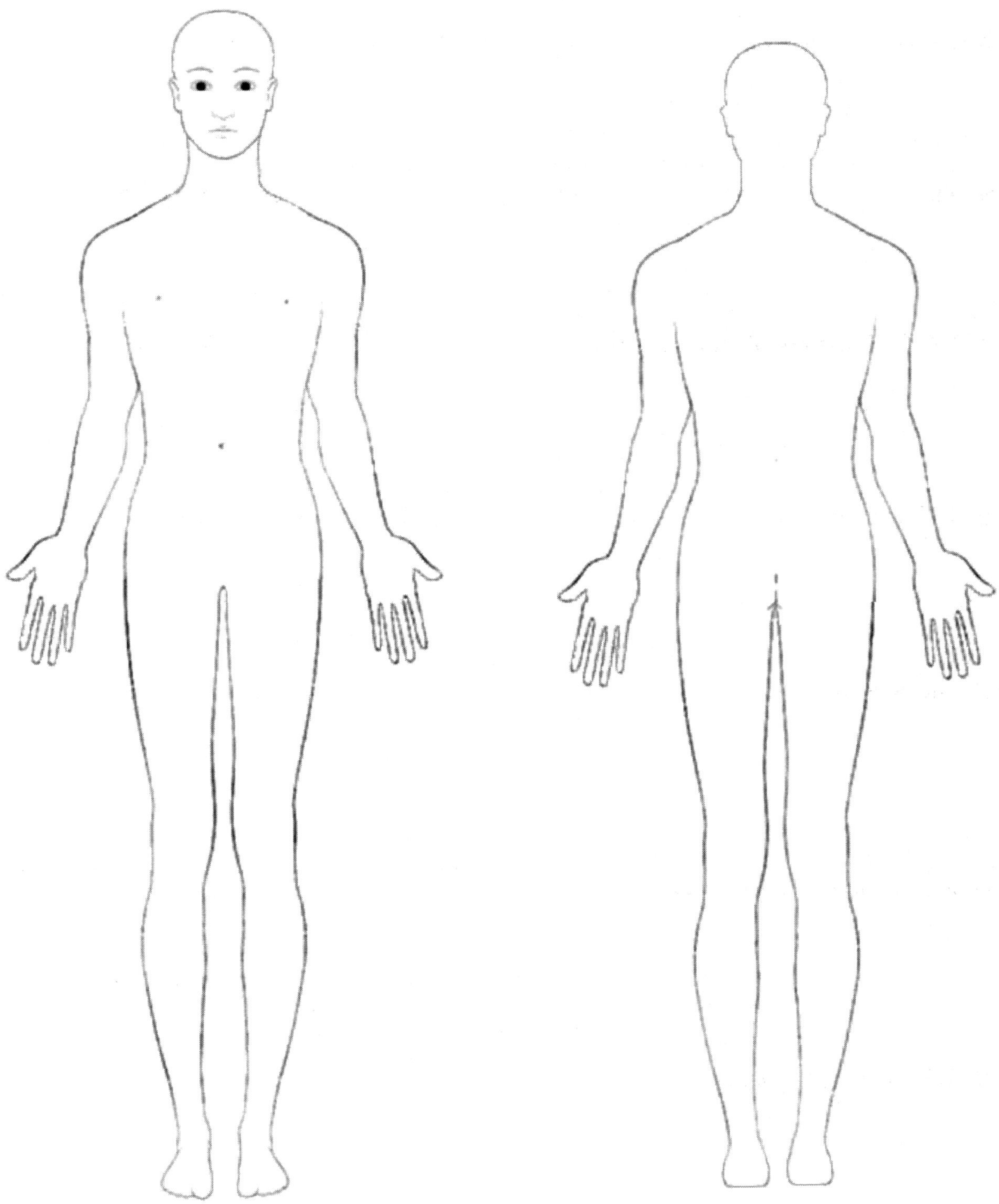

5. 캐릭터 분석 Character Analysis

<table>
<tr><th colspan="6">캐릭터 분석 Character Analysis</th></tr>
<tr><td colspan="3">작품명 Work Name</td><td colspan="2">장르 Genre</td><td>국가 Nationality</td></tr>
<tr><td colspan="3">배역명 Character Name</td><td colspan="3">역할 Part</td></tr>
<tr><td>성별 Sex</td><td colspan="3">나이 Age</td><td colspan="2">성격 Personality</td></tr>
<tr><td colspan="3">유전학적 특징 Genetic Characteristics</td><td colspan="3">시대적 배경 Historical Background</td></tr>
<tr><td colspan="3">건강 상태 Health Status</td><td colspan="3">생활 환경 Living Conditions</td></tr>
<tr><td colspan="6">특징 Characteristic</td></tr>
<tr><td colspan="6">성격 표현 Character Expression</td></tr>
<tr><td colspan="6">분장 의도 Makeup Purpose</td></tr>
</table>

6. 분장 계획표 Makeup Planning Chart

분장 계획표 Makeup Planning Chart

작품 명 Work Name	배역 명 Character Name
	배우 명 Actor's Name

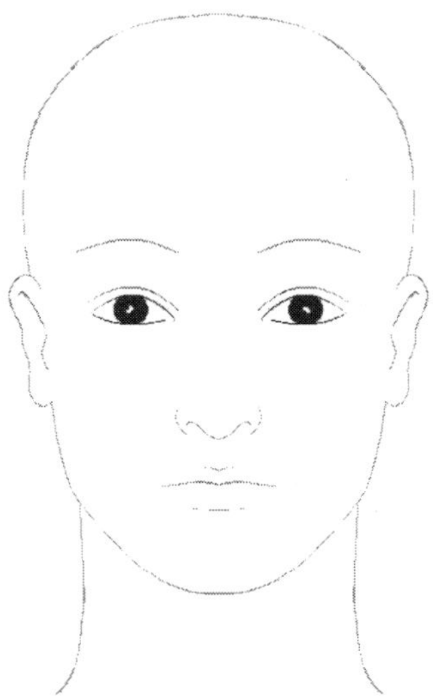

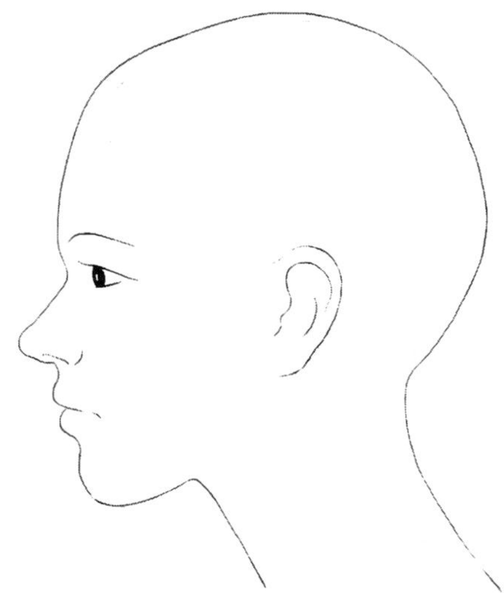

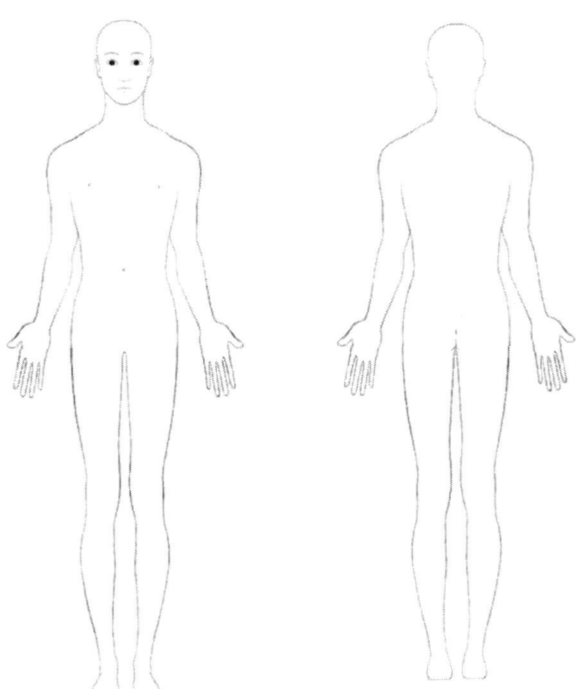

기초 Base	화운데이션 Foundation	음영 Contour
파우더 Powder	이마 Forehead	눈썹 Eyebrows
아이섀도 Eye shadow	아이라이너 Eye liners	마스카라 Mascara
코 Nose	뺨 Cheek	입 Mouth
헤어 Hair	수염 Beard	메모 Note

7. 캐릭터 분장 완성 Character Makeup Complete

분장 완성 이미지 삽입
Attached is an image of the Makeup completion.

Lesson 7

캐릭터 디자인 3
Character Design 3

Lesson 7. 캐릭터 디자인3 Character Design 3

1. 캐릭터 자료 조사 Character Research

이미지 삽입
Insert Image

2. 캐릭터 디자인 모티브 Character Design Motif

•

•

모티브 이미지 삽입
Insert Motif Image

모티브 이미지 삽입
Insert Motif Image

•

•

모티브 이미지 삽입
Insert Motif Image

모티브 이미지 삽입
Insert Motif Image

3. 분장 디자인 러프 스케치 Makeup Design Rough Sketch

1안 정면 1Draft Front View

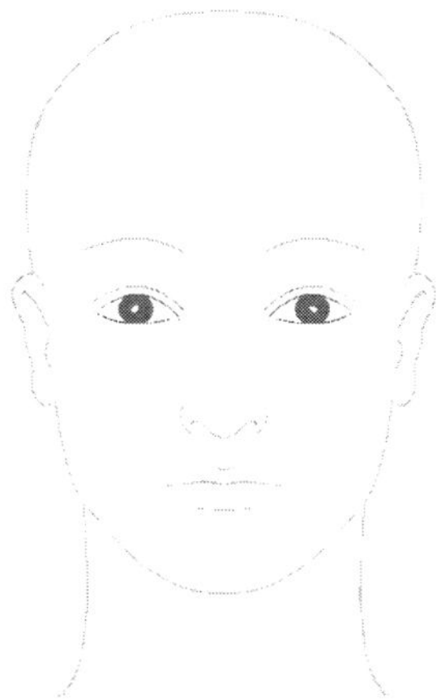

2안 정면 2Draft Front View

1안 측면 1 Draft Side View

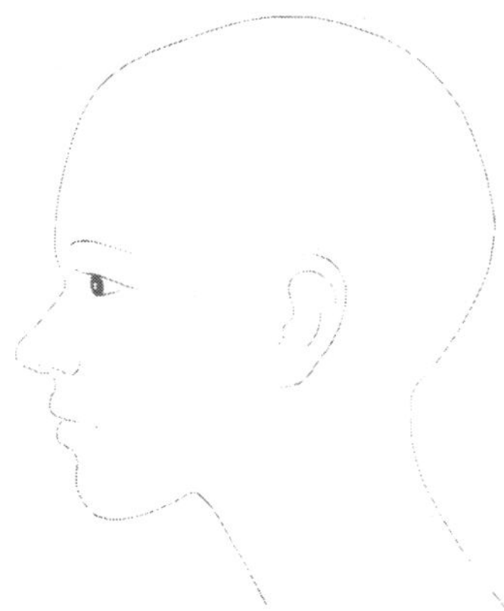

2안 측면 2 Draft Side View

전신 1 Full Body 1

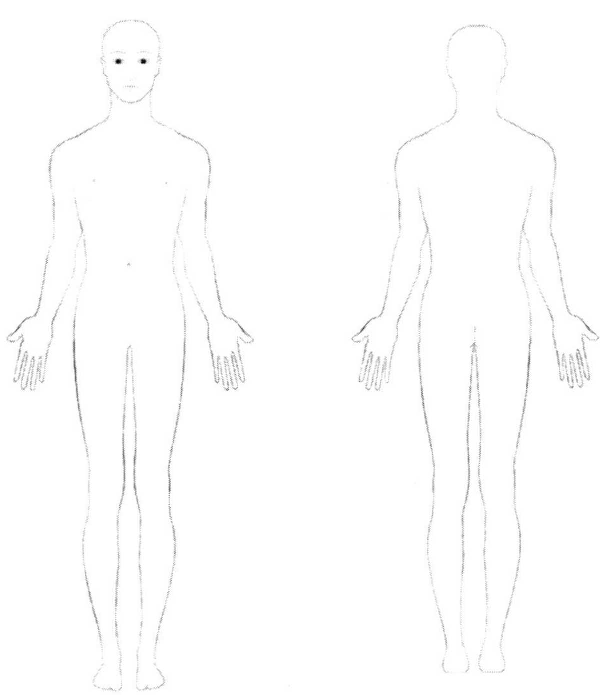

기타 Miscellaneous

4. 분장 디자인 - 정면 Makeup Design - Front

캐릭터의 정면 얼굴을 디자인하여 일러스트레이션하시오.

Design the front face of the character and create an illustration.

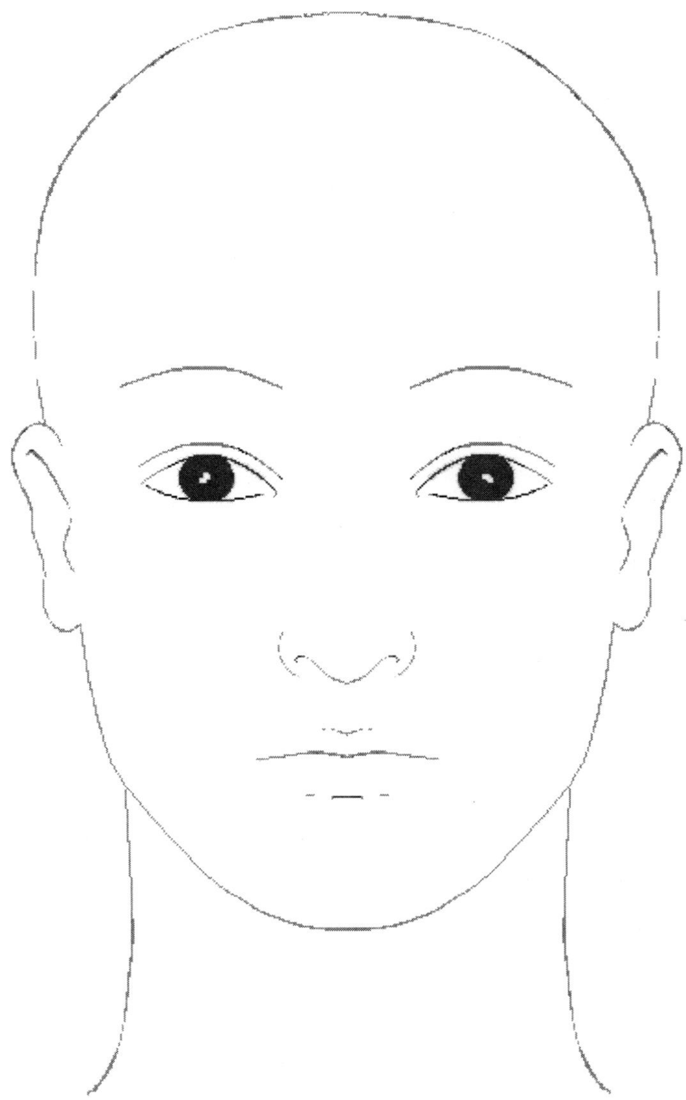

4. 분장 디자인 - 측면 Makeup Design - Side

캐릭터의 측면 얼굴을 디자인하여 일러스트레이션하시오.

Design and illustrate the side face of the character.

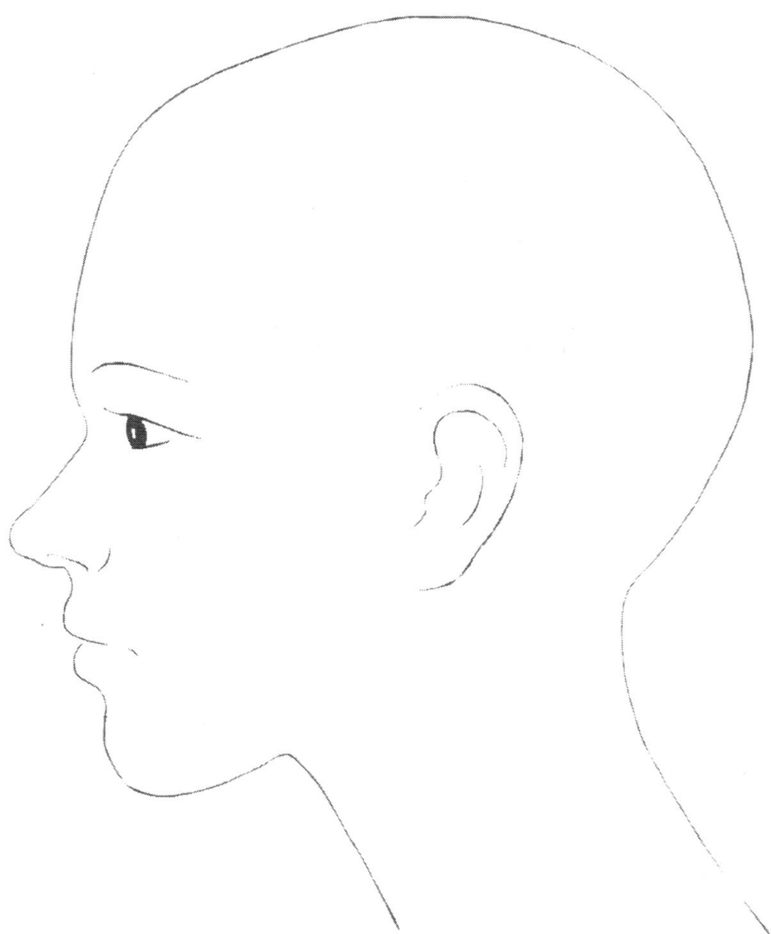

4. 분장 디자인 - 전신 Makeup Design - Full Body

캐릭터의 측면 얼굴을 디자인하여 일러스트레이션하시오.

Design and illustrate the side face of the character.

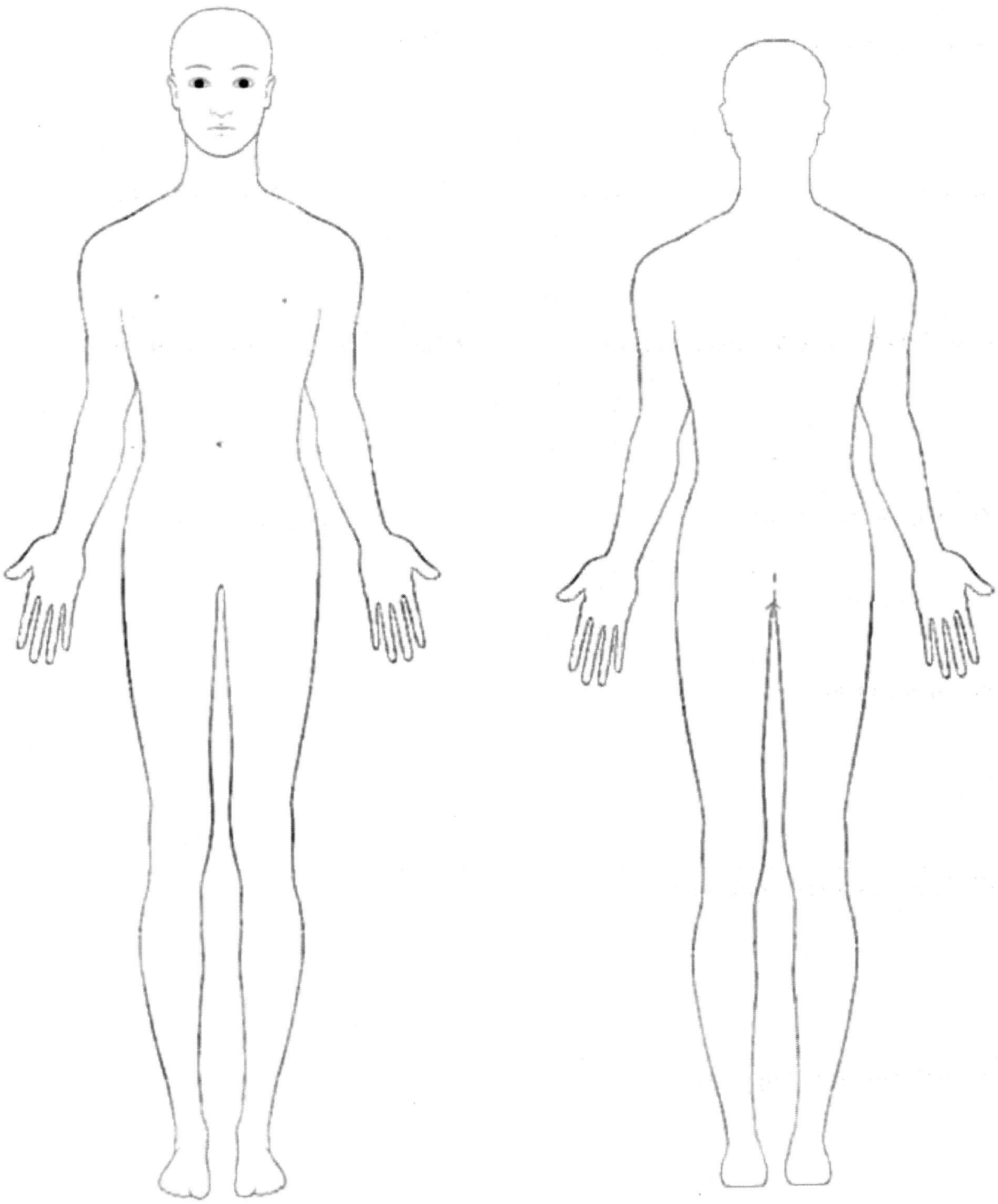

5. 캐릭터 분석 Character Analysis

<table>
<tr><th colspan="6">캐릭터 분석 Character Analysis</th></tr>
<tr><td colspan="3">작품명 Work Name</td><td colspan="2">장르 Genre</td><td>국가 Nationality</td></tr>
<tr><td colspan="3">배역명 Character Name</td><td colspan="3">역할 Part</td></tr>
<tr><td colspan="2">성별 Sex</td><td colspan="2">나이 Age</td><td colspan="2">성격 Personality</td></tr>
<tr><td colspan="3">유전학적 특징 Genetic Characteristics</td><td colspan="3">시대적 배경 Historical Background</td></tr>
<tr><td colspan="3">건강 상태 Health Status</td><td colspan="3">생활 환경 Living Conditions</td></tr>
<tr><td colspan="6">특징 Characteristic</td></tr>
<tr><td colspan="6">성격 표현 Character Expression</td></tr>
<tr><td colspan="6">분장 의도 Makeup Purpose</td></tr>
</table>

6. 분장 계획표 Makeup Planning Chart

분장 계획표 Makeup Planning Chart

작품 명 Work Name	배역 명 Character Name
	배우 명 Actor's Name

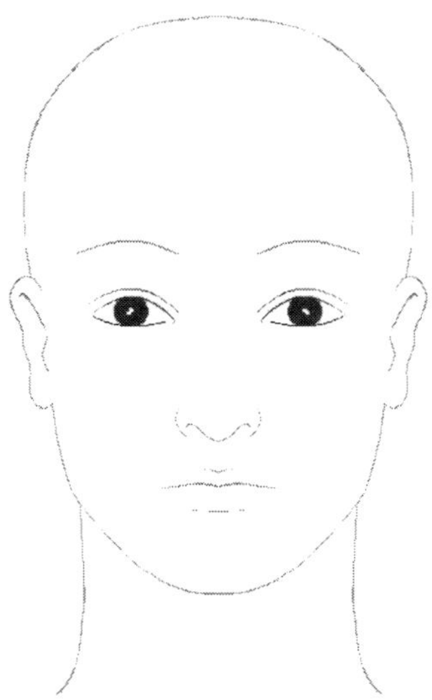

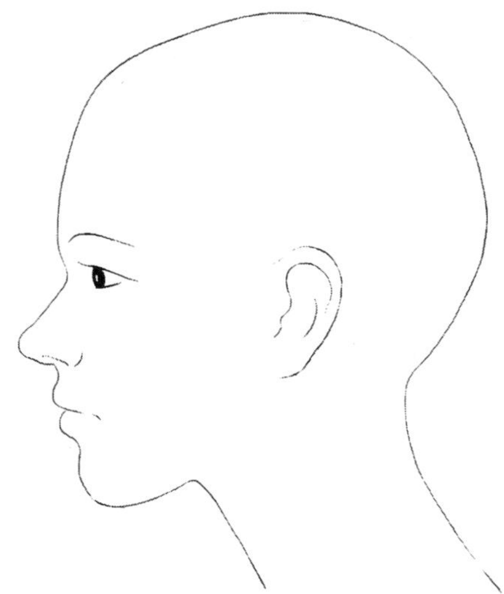

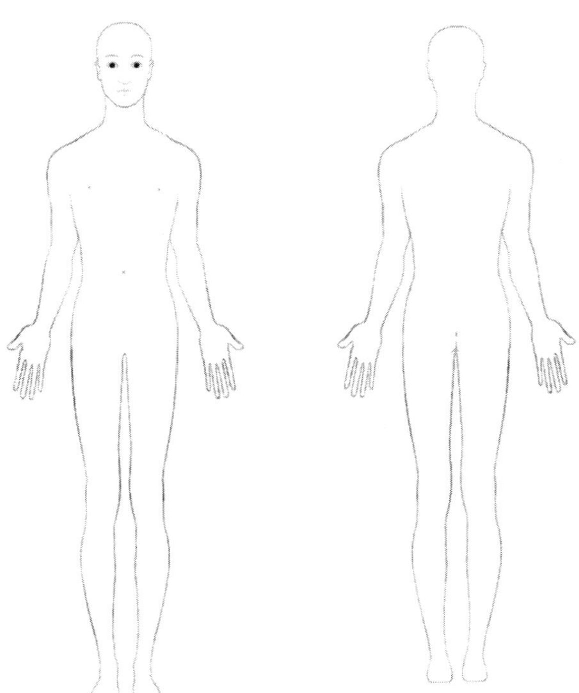

기초 Base	화운데이션 Foundation	음영 Contour
파우더 Powder	이마 Forehead	눈썹 Eyebrows
아이섀도 Eye shadow	아이라이너 Eye liners	마스카라 Mascara
코 Nose	뺨 Cheek	입 Mouth
헤어 Hair	수염 Beard	메모 Note

7. 캐릭터 분장 완성 Character Makeup Complete

분장 완성 이미지 삽입

Attached is an image of the Makeup completion.

Lesson 8

라이프 캐스팅
Life Casting

Lesson 8. 라이프 캐스팅 Life Casting

1. 자료조사 Data Research

2. 주의 사항 Precautions

-
-
-
-
-
-

3. 준비 작업 Preparation Work

■ 도구 사용 Tool Use

도구 이미지 삽입 Tool Image Insertion	도구 이미지 삽입 Tool Image Insertion
도구 이미지 삽입 Tool Image Insertion	도구 이미지 삽입 Tool Image Insertion
도구 이미지 삽입 Tool Image Insertion	도구 이미지 삽입 Tool Image Insertion
도구 이미지 삽입 Tool Image Insertion	도구 이미지 삽입 Tool Image Insertion
도구 이미지 삽입 Tool Image Insertion	도구 이미지 삽입 Tool Image Insertion

4. 도구 사용 Tool Use

도구 Tools	사용방법 How To Use
도구 이미지 삽입 Tool Image Insertion	
도구 이미지 삽입 Tool Image Insertion	
도구 이미지 삽입 Tool Image Insertion	
도구 이미지 삽입 Tool Image Insertion	
도구 이미지 삽입 Tool Image Insertion	
도구 이미지 삽입 Tool Image Insertion	

5. 볼드캡 Bold Cap

1) 볼드캡 제작 준비물 Preparations For Making Bold Caps

볼드캡 제작 준비물 이미지 삽입
Inserting Images Of Preparations For Making Bold Caps

2) 볼드캡 제작 과정 Process Of Putting Bold Caps On

볼드캡 제작 과정
이미지 삽입
Process For Making Bold Caps
Insert Image

1단계 Step 1

볼드캡 제작 과정
이미지 삽입
Process For Making Bold Caps
Insert Image

2단계 Step 2

볼드캡 제작 과정
이미지 삽입
Process For Making Bold Caps
Insert Image

3단계 Step 3

볼드캡 제작 과정
이미지 삽입
Process For Making Bold Caps
Insert Image

4단계 Step 4

볼드캡 제작 과정
이미지 삽입
Process For Making Bold Caps
Insert Image

5단계 Step 5

볼드캡 제작 과정
이미지 삽입
Process For Making Bold Caps
Insert Image

6단계 Step 6

3) 볼드캡 씌우기 작업 과정 Process Of Putting Bold Caps On

볼드캡 적용 과정
이미지 삽입
Insert Image Of Bold Cap
Application Process

1단계 Step 1

볼드캡 적용 과정
이미지 삽입
Insert Image Of Bold Cap
Application Process

2단계 Step 2

볼드캡 적용 과정
이미지 삽입
Insert Image Of Bold Cap
Application Process

3단계 Step 3

볼드캡 적용 과정
이미지 삽입
Insert Image Of Bold Cap
Application Process

4단계 Step 4

볼드캡 적용 과정
이미지 삽입
Insert Image Of Bold Cap
Application Process

5단계 Step 5

볼드캡 적용 과정
이미지 삽입
Insert Image Of Bold Cap
Application Process

6단계 Step 6

4) 볼드캡 제거 작업 과정 Process Of Removing Bold Caps

볼드캡 제거 과정
이미지 삽입

1단계 Step 1

볼드캡 제거 과정
이미지 삽입

2단계 Step 2

볼드캡 제거 과정
이미지 삽입

3단계 Step 3

6. 알지네이트 Alginate

1) 알지네이트 Alginate

- 치과에서 사용되는 재료로 인체에 무해하여 모형을 뜰때 안전한 재료.

알지네이트 이미지 삽입
Insert Alginate Image

2) 알지네이트 작업 과정 Alginate Process

알지네이트 작업 과정
이미지 삽입
Insert Image Of Alginate Process

1단계 Step

알지네이트 작업 과정
이미지 삽입
Insert Image Of Alginate Process

2단계 Step

알지네이트 작업 과정
이미지 삽입
Insert Image Of Alginate Process

3단계 Step

알지네이트 작업 과정
이미지 삽입
Insert Image Of Alginate Process

4단계 Step

알지네이트 작업 과정
이미지 삽입
Insert Image Of Alginate Process

5단계 Step

알지네이트 작업 과정
이미지 삽입
Insert Image Of Alginate Process

6단계 Step

7. 석고붕대 Plaster Bandage

1) 석고붕대 Plaster Bandage

- 라이프 캐스팅 작업 시 알지네이트을 보호하기 위해 사용
-
-
-
-

석고 붕대 이미지 삽입
Attached plaster bandage image

2) 석고붕대 작업 과정 Plaster Bandage Work Process

석고 붕대 작업 과정
이미지 삽입
Plaster Bandage Work Process
Image Insertion

1단계 Step 1

석고 붕대 작업 과정
이미지 삽입
Plaster Bandage Work Process
Image Insertion

2단계 Step 2

석고 붕대 작업 과정
이미지 삽입
Plaster Bandage Work Process
Image Insertion

3단계 Step 3

석고 붕대 작업 과정
이미지 삽입
Plaster Bandage Work Process
Image Insertion

4단계 Step 4

석고 붕대 작업 과정
이미지 삽입
Plaster Bandage Work Process
Image Insertion

5단계 Step 5

석고 붕대 작업 과정
이미지 삽입
Plaster Bandage Work Process
Image Insertion

6단계 Step 6

8. 석고 Plaster

1) 석고의 종류 Types Of Plaster

•

석고 이미지 삽입
Insert Plaster Image

•

석고 이미지 삽입
Insert Plaster Image

•

석고 이미지 삽입
Insert Plaster Image

•

석고 이미지 삽입
Insert Plaster Image

9. 양각 몰드 작업 Positive Mold

1) 석고 Plaster

석고 이미지 삽입
Insert Plaster Image

2) 양각 몰드 작업 과정 Embossed Mold Work Process

양각 몰드 제작 과정 이미지 삽입
Insert Image Of Embossed Mold Making Process

1단계 Step 1

양각 몰드 제작 과정 이미지 삽입
Insert Image Of Embossed Mold Making Process

2단계 Step 2

양각 몰드 제작 과정 이미지 삽입
Insert Image Of Embossed Mold Making Process

3단계 Step 3

양각 몰드 제작 과정 이미지 삽입
Insert Image Of Embossed Mold Making Process

4단계 Step 4

양각 몰드 제작 과정 이미지 삽입
Insert Image Of Embossed Mold Making Process

5단계 Step 5

양각 몰드 제작 과정 이미지 삽입
Insert Image Of Embossed Mold Making Process

6단계 Step 6

Lesson 9

조소
Sculpting

Lesson 9. 조소 Sculpting

1. 수 도구 Hand Tools

1) 조소도구의 사용 Use of Sculpting Tools

- 조소도구는 작업 순서의 단계에 따라 도구의 사용이 달라진다.

2) 조소도구의 종류 Types of Sculpting Tools

☐ 조소용 헤라

조고 도구 이미지 삽입 Inserting Images of Sculpting Tools	•	조고 도구 이미지 삽입 Inserting Images of Sculpting Tools	•
조고 도구 이미지 삽입 Inserting Images of Sculpting Tools	•	조고 도구 이미지 삽입 Inserting Images of Sculpting Tools	•
조고 도구 이미지 삽입 Inserting Images of Sculpting Tools	•	조고 도구 이미지 삽입 Inserting Images of Sculpting Tools	•
조고 도구 이미지 삽입 Inserting Images of Sculpting Tools	•	조고 도구 이미지 삽입 Inserting Images of Sculpting Tools	•

☐ 조소용 툴- 형태돌출작업시 사용

조고 도구 이미지 삽입 Inserting Images of Sculpting Tools	•	조고 도구 이미지 삽입 Inserting Images of Sculpting Tools	•
조고 도구 이미지 삽입 Inserting Images of Sculpting Tools	•	조고 도구 이미지 삽입 Inserting Images of Sculpting Tools	•
조고 도구 이미지 삽입 Inserting Images of Sculpting Tools	•	조고 도구 이미지 삽입 Inserting Images of Sculpting Tools	•
조고 도구 이미지 삽입 Inserting Images of Sculpting Tools	•	조고 도구 이미지 삽입 Inserting Images of Sculpting Tools	•
조고 도구 이미지 삽입 Inserting Images of Sculpting Tools	•	조고 도구 이미지 삽입 Inserting Images of Sculpting Tools	•

□ 조소용 툴- 미세 작업시 사용

조고 도구 이미지 삽입
Inserting Images of Sculpting Tools

•

조고 도구 이미지 삽입
Inserting Images of Sculpting Tools

•

조고 도구 이미지 삽입
Inserting Images of Sculpting Tools

•

조고 도구 이미지 삽입
Inserting Images of Sculpting Tools

•

조고 도구 이미지 삽입
Inserting Images of Sculpting Tools

•

조고 도구 이미지 삽입
Inserting Images of Sculpting Tools

•

조고 도구 이미지 삽입
Inserting Images of Sculpting Tools

•

조고 도구 이미지 삽입
Inserting Images of Sculpting Tools

•

조고 도구 이미지 삽입
Inserting Images of Sculpting Tools

•

조고 도구 이미지 삽입
Inserting Images of Sculpting Tools

•

☐ 조소용 툴- 스킨 표현 작업시 사용

I스킨 표현 도구 이미지 삽입 Skin Expression Tool Image Insertion	•	스킨 표현 도구 이미지 삽입 Skin Expression Tool Image Insertion	•
스킨 표현 도구 이미지 삽입 Skin Expression Tool Image Insertion	•	스킨 표현 도구 이미지 삽입 Skin Expression Tool Image Insertion	•

3) 표면 정리 제품 Surface Preparation Products

표면 정리 도구 이미지 삽입 Inserting Images of Surface Preparation Tools	•	표면 정리 도구 이미지 삽입 Inserting Images of Surface Preparation Tools	•
표면 정리 도구 이미지 삽입 Inserting Images of Surface Preparation Tools	•	표면 정리 도구 이미지 삽입 Inserting Images of Surface Preparation Tools	•
표면 정리 도구 이미지 삽입 Inserting Images of Surface Preparation Tools	•	표면 정리 도구 이미지 삽입 Inserting Images of Surface Preparation Tools	•

2. 공구 Power Tools

도구 이미지 삽입
Inserting Tool Images

•

도구 이미지 삽입
Inserting Tool Images

•

도구 이미지 삽입
Inserting Tool Images

•

도구 이미지 삽입
Inserting Tool Images

•

도구 이미지 삽입
Inserting Tool Images

•

도구 이미지 삽입
Inserting Tool Images

•

도구 이미지 삽입
Inserting Tool Images

•

도구 이미지 삽입
Inserting Tool Images

•

도구 이미지 삽입
Inserting Tool Images

•

도구 이미지 삽입
Inserting Tool Images

•

3. 재료 Materials

1) 일반 유토-유황성분 있음 General Oil - Contains Sulfur

- 원형 제작용 유토로 지속적으로 사용 가능한 oil Based Clay 유토
- 유토의 단단한 정도에 따른 차이가 있음
- 핫폼과 콜드 폼에 사용 가능하나 실리콘 사용은 불가
- 보관 통에 담아 보관

이미지 삽입 Insert Image	이미지 삽입 Insert Image	이미지 삽입 Insert Image

2) 샤반트 유토- 유황성분 없음 Chavant NSP- Does Not Contain Sulfur

- 원형 제작용조형용 유토로 지속적으로 사용 가능한 모델링 oil Based Clay 유토
- 유토의 단단한 정도에 따른 차이가 있음
- 무독성 제품으로 안전 하게 사용 가능ASTM: 미국 재료시험 현회 인증
- 실리콘, 핫폼, 콜드 폼 사용 가능
- 보관 통에 담아 보관

이미지 삽입 Insert Image	이미지 삽입 Insert Image	이미지 삽입 Insert Image
부드러운 유토 Soft Clay	**미디움 유토 Medium Clay**	**단단한 유토 Hard Clay**

3) 찰흙 Clay

-
-

이미지 삽입
Insert Image

4. 형태제작 Shape Making

•

정면 이미지 삽입
Insert Front View Image

정면이미지
Front View Image

뒷면 이미지 삽입
Insert back view image

뒷면이미지
Back View Image

좌측면 이미지 삽입
Insert left side view image

좌측면 이미지
Left Side View Image

우측면 이미지 삽입
Insert Right Side View Image

우측면 이미지
Right Side ViewImage

윗면 이미지 삽입
Insert Top View Image

윗면 이미지
Top View Image

5. 조소 비율 Sculpture Proportions

•

정면 이미지 삽입
Insert Front View Image

정면이미지
Front View Image

뒷면 이미지 삽입
Insert back view image

뒷면이미지
Back View Image

좌측면 이미지 삽입
Insert left side view image

좌측면 이미지
Left Side View Image

우측면 이미지 삽입
Insert Right Side View Image

우측면 이미지
Right Side ViewImage

윗면 이미지 삽입
Insert Top View Image

윗면 이미지
Top View Image

6. 세부 과정 Detailed Process

•

정면 이미지 삽입 Insert Front View Image	뒷면 이미지 삽입 Insert back view image
정면이미지 Front View Image	**뒷면이미지** Back View Image

좌측면 이미지 삽입 Insert left side view image	우측면 이미지 삽입 Insert Right Side View Image	윗면 이미지 삽입 Insert Top View Image
좌측면 이미지 Left Side View Image	**우측면 이미지** Right Side ViewImage	**윗면 이미지** Top View Image

7. 조소 작업 완성 Sculpture Work Completed

•

정면 이미지 삽입 Insert Front View Image	뒷면 이미지 삽입 Insert back view image
정면이미지 Front View Image	**뒷면이미지** Back View Image

좌측면 이미지 삽입 Insert left side view image	우측면 이미지 삽입 Insert Right Side View Image	윗면 이미지 삽입 Insert Top View Image
좌측면 이미지 Left Side View Image	**우측면 이미지** Right Side ViewImage	**윗면 이미지** Top View Image

Lesson 10

후반작업
Post-Production

Lesson 10. 후반작업 Post-Production

1. 네가티브 몰드 Negative Mold

1) 네가티브 몰드 제작 재료 Negative Mold Making Materials

■ 석고 붕대 / 가루석고 Plaster bandage / Plaster

재료 이미지 삽입
Insert materials image

■ 찰흙 Clay

재료 이미지 삽입
Insert materials image

■ 바세린 Vaseline

재료 이미지 삽입
Insert materials image

■ 용기 Container

재료 이미지 삽입
Insert materials image

■ 기타 Miscellaneous

재료 이미지 삽입
Insert materials image

2) 재료 및 도구 Materials and tools

■ 네가티브 몰드 제작에 사용되는 점토 도구 Clay tools used in making negative molds

•	조소 도구 이미지 삽입 Insert an image of sculpting tools	•	조소 도구 이미지 삽입 Insert an image of sculpting tools
•	조소 도구 이미지 삽입 Insert an image of sculpting tools	•	조소 도구 이미지 삽입 Insert an image of sculpting tools
•	조소 도구 이미지 삽입 Insert an image of sculpting tools	•	조소 도구 이미지 삽입 Insert an image of sculpting tools
•	조소 도구 이미지 삽입 Insert an image of sculpting tools	•	조소 도구 이미지 삽입 Insert an image of sculpting tools
•	조소 도구 이미지 삽입 Insert an image of sculpting tools	•	조소 도구 이미지 삽입 Insert an image of sculpting tools
	조소 도구 이미지 삽입 Insert an image of sculpting tools		조소 도구 이미지 삽입 Insert an image of sculpting tools

3) 석고 네가티브 몰드 작업 과정 Plaster Negative Mold Process

이미지 삽입 Insert Image.	이미지 삽입 Insert Image.	이미지 삽입 Insert Image.
1단계: 원본 준비 Step1: Preparation of the Original Object	2단계: 몰드 박스 제작 Step2: Building a Mold Box	3단계: 용기에 물을 채워 준비 Step3: Prepare to fill a container with water
이미지 삽입 Insert Image.	이미지 삽입 Insert Image.	이미지 삽입 Insert Image.
4단계: 석고 가루 넣기 Step4: Adding plaster powder	5단계: 석고 혼합 Step5: Mixing the Plaster	6단계: 석고 붓기 Step6: Pouring the Plaster
이미지 삽입 Insert Image.	이미지 삽입 Insert Image.	이미지 삽입 Insert Image.
7단계: 석고 굳히기 Step7: Curing the Plaster:	8단계: 네거티브 몰드 분리 Step8: Removing the Negative Mold	9단계: 마무리 작업 Step9: Finishing Touches

4) 조소 비율 Sculptural proportions

정면 이미지 삽입 Insert Front View Image	뒷면 이미지 삽입 Insert back view image
정면 Front View	뒷면 Back View

좌측면 이미지 삽입 Insert left side view image	우측면 이미지 삽입 Insert Right Side View Image	윗면 이미지 삽입 Insert Top View Image
좌측면 Left Side View	우측면 Right Side View	윗면 Top View

5) 믹싱 작업 Mixing Operations

이미지 삽입
Image attachment

6) 오븐 작업 Oven Operations

이미지 삽입
Insert Image.

7) 수세 및 건조 Washing And Drying

이미지 삽입
Insert Image.

8) 트리밍 Trimming

□

□

이미지 삽입
Insert Image.

2. 후반 작업 Post-Production

1) 재료 및 도구 Materials And Tools

□ 네가티브 몰드 제작에 사용되는 재료의 종류 Types Of Materials Used In Making Negative Molds

•	이미지 삽입 Insert Image.	•	이미지 삽입 Insert Image.
•	이미지 삽입 Insert Image.	•	이미지 삽입 Insert Image.
•	이미지 삽입 Insert Image.	•	이미지 삽입 Insert Image.
•	이미지 삽입 Insert Image.	•	이미지 삽입 Insert Image.
•	이미지 삽입 Insert Image.	•	이미지 삽입 Insert Image.

2) 후반 작업에 사용되는 재료 Materials Used In Post-Production

■ 콜드폼 Cold Foam

① 폴리우레탄 폼 PU Foam:

이미지 삽입
Insert Image.

② 촉진제 Catalyst:

이미지 삽입
Insert Image.

③ 발포제 Blowing agent

이미지 삽입
Insert Image.

④ 경화제 Hardener:

이미지 삽입
Insert Image.

⑤ 착색제 Pigment:

이미지 삽입
Insert Image.

⑥ 분리제 Separator

■ 핫 폼 Hot Foam

① 라텍스 베이스 Latex Base

이미지 삽입
Insert Image.

② 발포제 Foaming Agent

이미지 삽입
Insert Image.

③ 경화제 Curing Agent

이미지 삽입
Insert Image.

④ 겔화 에이전트 Gelling Agent

이미지 삽입
Insert Image.

⑤ 몰드 릴리스 Mold Release

이미지 삽입
Insert Image.

■ 실리콘 Silicone

①

실리콘 제품 이미지 삽입
Insert silicone product image

실리콘 제품 이미지 삽입
Insert silicone product image

②

실리콘 제품 이미지 삽입
Insert silicone product image

실리콘 제품 이미지 삽입
Insert silicone product image

③

실리콘 제품 이미지 삽입
Insert silicone product image

■ 채색용품 Coloring Supplies

•

제품 이미지 삽입
Insert product image

•

제품 이미지 삽입
Insert product image

•

제품 이미지 삽입
Insert product image

•

제품 이미지 삽입
Insert product image

•

제품 이미지 삽입
Insert product image

Lesson 11

특수 분장 적용
Apply Special Effects Makeup

Lesson 11. 특수 분장 적용 Apply Special Effects Makeup

1. 특수 분장 적용 재료 Special Makeup Application Materials

1) 핫 폼 Hot foam

이미지 삽입
Insert Image

2) 콜드폼 Cold foam

이미지 삽입
Insert Image

3) 실리콘 Silicone

이미지 삽입
Insert Image

4) 채색용품 Coloring Supplies

이미지 삽입
Insert Image

2. 분장 적용 Makeup Application

정면 이미지 삽입
Insert Front View Image

정면이미지
Front View Image

뒷면 이미지 삽입
Insert back view image

뒷면이미지
Back View Image

좌측면 이미지 삽입
Insert left side view image

좌측면 이미지
Left Side View Image

우측면 이미지 삽입
Insert Right Side View Image

우측면 이미지
Right Side ViewImage

윗면 이미지 삽입
Insert Top View Image

윗면 이미지
Top View Image

3. 채색 Coloring

정면 이미지 삽입
Insert Front View Image

정면이미지
Front View Image

뒷면 이미지 삽입
Insert back view image

뒷면이미지
Back View Image

좌측면 이미지 삽입
Insert left side view image

좌측면 이미지
Left Side View Image

우측면 이미지 삽입
Insert Right Side View Image

우측면 이미지
Right Side ViewImage

윗면 이미지 삽입
Insert Top View Image

윗면 이미지
Top View Image

4. 분장 디자인 적용 부위 Make-Up Design Application Area

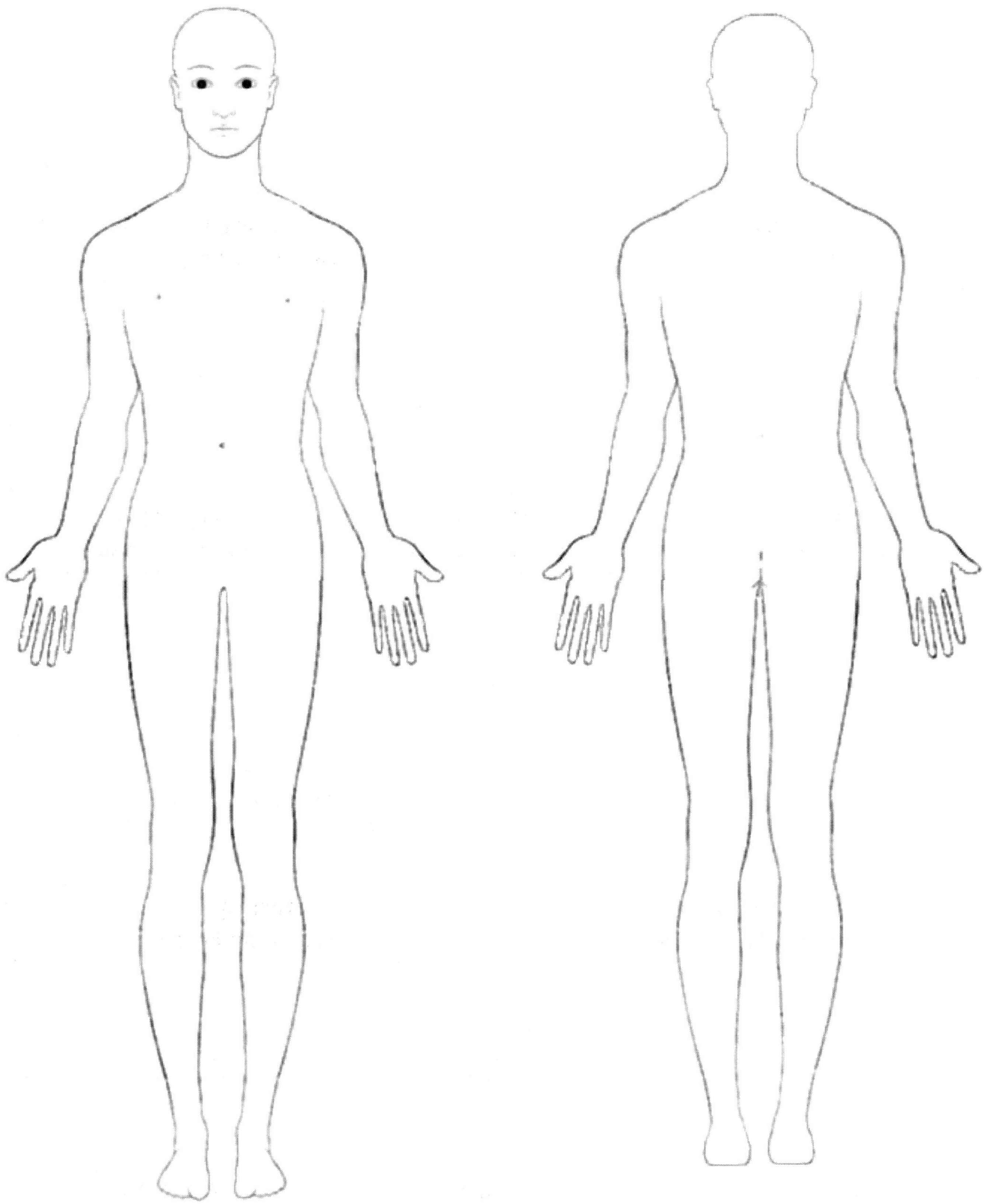

5. 소품 Props

이미지 삽입
Insert Image

이미지 삽입
Insert Image

이미지 삽입
Insert Image

6. 특수 분장 완성 Special Makeup Completion

정면 이미지 삽입
Insert Front View Image

정면이미지Front View Image

뒷면 이미지 삽입
Insert back view image

뒷면이미지Back View Image

좌측면 이미지 삽입
Insert left side view image

좌측면 이미지Left Side View Image

우측면 이미지 삽입
Insert Right Side View Image

우측면 이미지Right Side ViewImage

7. 분장 계획표 Makeup Plan

<table>
<tr><th colspan="5">분장 계획표 Makeup Plan</th></tr>
<tr><td>Work Name
작품명</td><td colspan="2"></td><td>Actor
배우</td><td></td></tr>
<tr><td colspan="5">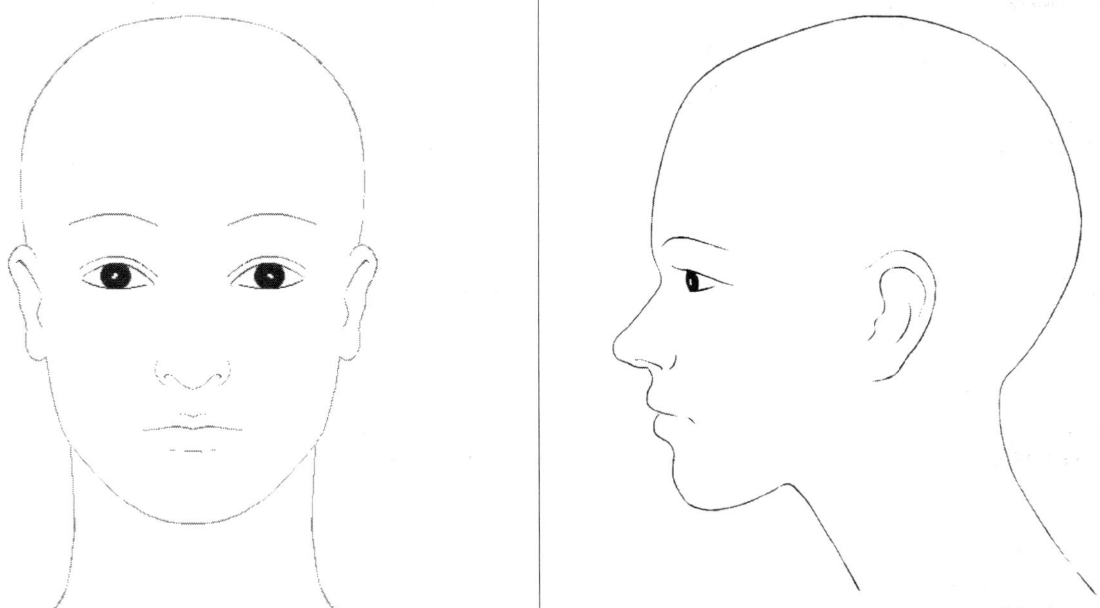</td></tr>
<tr><td>이마 Forehead</td><td>눈 Eye</td><td>코 Nose</td><td>뺨 Cheek</td><td>입 Mouth</td></tr>
<tr><td>눈썹 Eyebrows</td><td>화운데이션 Foundation</td><td>밝은부분 Highlight</td><td>명암 Shading</td><td>파우더 Powder</td></tr>
<tr><td>목 Neck</td><td>팔 Arm</td><td>다리 Leg</td><td>복부 Stomach</td><td>헤어 Hair</td></tr>
<tr><td>어깨 Shoulder</td><td>손 Hand</td><td>발 Foot</td><td>등 Back</td><td>메모 Note</td></tr>
</table>

8. 캐릭터 분석 및 분장 의도 Character Analysis And Makeup Intention

1) 캐릭터 분석 Character Analysis

캐릭터 분석 Character Analysis	
작품명 Work Name	장르 Genre
배역명 Character Name	역할 Part
나이/성별 Age/Sex	성격 Personality
건강 Health	환경 Environment
유전학 Genetics	시대/국적 Times/Nationality
특징 Characteristic	

2) 분장의도 Makeup Purpose

설명 Explain	
분장 의도 Makeup Purpose	
성격 표현 Character Expression	

Lesson 12

스케일 모델
Scale Model

Lesson 12 스케일 모델 Scale Model

1. 스케일 모델이란? What is a scale model?

1) 비례 표현 Proportional Expressioncaling

2) 스케일링 Scaling

1:6	
1:12	
1:50	
1:100	

2. 자료조사 Data Research

3. 재료 Materials

■ **유토 Clay**

이미지 삽입
Image attachment

■ **찰흙 Clay**

이미지 삽입
Image attachment

■

이미지 삽입
Image attachment

■

이미지 삽입
Image attachment

4. 부분제작 Partial Production

■ 뿔 Horns

이미지 삽입
Image attachment

■ 귀 Ears

이미지 삽입
Image attachment

■ 치아 Teeth

이미지 삽입
Image attachment

■ 안구 Eyes

이미지 삽입
Image attachment

5. 소품 Props

■

이미지 삽입
Image attachment

■

이미지 삽입
Image attachment

■

이미지 삽입
Image attachment

■

이미지 삽입
Image attachment

6. 스케일 모델 디자인 - 정면 Scale Model Design - Front

스케일 모델 디자인 정면을 일러스트 하세요.
Illustration of the Front of the Scale Model Design

6. 스케일 모델 디자인 - 측면 Scale Model Design - Side View

스케일 모델 디자인 측면을 일러스트 하세요.
Illustration of the Side View of a Scale Model Design

6. 스케일 모델 디자인 - 전신 Scale Model Design - Full Body

스케일 모델 디자인 전신을 일러스트 하세요.
Scale Model Design Illustration of the Whole Body

7. 스케일 모델 완성 Scale Model Completion

정면 이미지 삽입 Insert Front View Image	뒷면 이미지 삽입 Insert back view image
정면 이미지 Front View Image	**뒷면 이미지 Back View Image**
좌측면 이미지 삽입 Insert left side view image	우측면 이미지 삽입 Insert Right Side View Image
좌측면 이미지 Left Side View Image	**우측면 이미지 Right Side ViewImage**

8. 스케일링 모델 분석 Scaling Model Analysis

캐릭터 분석 Character Analysis	
작품명 Work Name	장르 Genre
스케일링 모젤 축소 비율	역할 Part
나이/성별 Age/Sex	성격 Personality
건강 Health	환경 Environment
유전학 Genetics	시대/국적 Times/Nationality
특징 Characteristic	
스케일링 모델 제작 의도 Scaling Model Production Intention	
스케일링 모델 표현 포인트 Scaling Model Expression Points	

Lesson 13

특수분장 제작 기록

Special Makeup Production Records

Lesson 13. 특수분장 제작 기록 Special Makeup Production Records

1. 인트로 Introduction

☐ 제목 Title

☐ 내용 Content

2. 디자인 컨셉 Design Concept

☐ 내용 Content

☐ 포인트 Points

이미지 삽입
Insert Image

3. 재료 Materials

□ 내용 Content

□ 포인트 Points

이미지 삽입
Insert Image

4. 조형물 기초 Sculpture Basics

□ 내용 Content

□ 포인트 Points

이미지 삽입
Insert Image

5. 질감 표현 Texture Expression

☐ 내용 Content

☐ 포인트 Points

이미지 삽입
Insert Image

6. 채색 Coloring

☐ 내용 Content

☐ 포인트 Points

이미지 삽입
Insert Image

7. 소품 Props

☐ 내용 Content

☐ 포인트 Points

이미지 삽입
Insert Image

8. 최종 연출 Final Touches and Presentation

☐ 내용 Content

☐ 포인트 Points

이미지 삽입
Insert Image

9. 엔딩 Conclusion

☐ 내용 Content

☐ 포인트 Points

이미지 삽입
Insert Image

10. 에필로그 Epilogue

☐ 내용 Content

☐ 포인트 Points

이미지 삽입
Insert Image

참고 도서

- 김정필(2010). 조형재료학. 재원미술총서.
- 윤예령, 김지연, 이선주, 임희경, 최정아(2022). 미디어 메이크업. 메이시언.
- 임희경, 신세영(2019). Special effects Make-up. 청구문화사.
- 임희경, 석현정, HUO TAO.(2020). Media Make-up(미디어 메이크업). 청구문화사.
- Richard Corson.(2000). Stage Makeup. Taylor & Francis; Subsequent edition
- Vincent Kehoe.(1991) Special Make-Up Effects. Routledge; 1st edition
- Todd Debreceni.(2018). Special Makeup Effects for Stage and Screen: Making and Applying Prosthetics. Routledge; 3rd edition
- Christopher Payne .(2021). A Beginner's Guide to Special Makeup Effects: Monsters, Maniacs and More 1st Edition. Routledge; 1st edition.

분장 관련 전문 업체

- 메이크업 매직 .http://m.makeupmagic.co.kr
- 협진 물산. https://www.hyup-shin.co.kr
- bennye, https://www.bennye.com/
- kryolan. https://kryolanshop.com/
- mehron. https://www.mehron.com/
- smooth-on. https://www.smooth-on.com/

Special Effects Makeup

2019년 3월 10일 초판 발행
2024년 10월 30일 2판 발행

지은이 : 임희경, HUO TAO
펴낸이 : 이상태
펴낸곳 : 청구문화사
주 소 : 03706 서울특별시 서대문구 성산로11길 15
전 화 : 02 332-0541代 FAX : 02 337-9577
등 록 : 1990년 11월 2일 제9-247호

정가 18,000원

ISBN 978-89-5616-944-6 93590